Understanding Urban Cycling

Academic interest in cycling has burgeoned in recent years with significant literature relating to the health and environmental benefits of cycling, the necessity for cycle-specific infrastructure, and the embodied experiences of cycling.

Based upon primary research in a variety of contexts such as London, Shanghai and Taipei, this book demonstrates that recent developments in urban cycling policy and practice are closely linked to broader processes of capital accumulation. It argues that cycling is increasingly caught up in discourses around smart cities that emphasise technological solutions to environmental problems and neoliberal ideas on individual responsibility and bio-political conduct, which only results in solutions that prioritise those who are already mobile. Accordingly, the central argument of the book is not that the popularisation of cycling is inherently bad, but that the manner in which cycling is being popularised gives cause for social and environmental concern. Ultimately the book argues that cycling has now become a vehicle for sustaining pro-growth agendas rather than subverting them or shifting to sustainable no-growth/de-growth and less technologically driven visions of modernity.

This book makes an innovative contribution to the fields of Cycling Studies, Mobilities and Transport and will be of interest to students and academics working in Human Geography, Transport Studies, Urban Studies, Urban Planning, Public Policy, Sociology and Sustainability.

Justin Spinney is Lecturer in Human Geography at the School of Geography and Planning at Cardiff University, UK.

Routledge Equity, Justice and the Sustainable City series

Series editors: Julian Agyeman and Stephen Zavestoski

This series positions equity and justice as central elements of the transition toward sustainable cities. The series introduces critical perspectives and new approaches to the practice and theory of urban planning and policy that ask how the world's cities can become 'greener' while becoming more fair, equitable and just.

Routledge Equity, Justice and the Sustainable City series addresses sustainable city trends in the global North and South and investigates them for their potential to ensure a transition to urban sustainability that is equitable and just for all. These trends include municipal climate action plans; resource scarcity as tipping points into a vortex of urban dysfunction; inclusive urbanisation; 'complete streets' as a tool for realising more 'liveable cities'; the use of information and analytics toward the creation of 'smart cities'.

The series welcomes submissions for high-level cutting-edge research books that push thinking about sustainability, cities, justice and equity in new directions by challenging current conceptualisations and developing new ones. The series offers theoretical, methodological, and empirical advances that can be used by professionals and as supplementary reading in courses in urban geography, urban sociology, urban policy, environment and sustainability, development studies, planning, and a wide range of academic disciplines.

Urban Gardening as Politics
Edited by Chiara Tornaghi and Chiara Certomà

Disruptive Transport
Driverless Cars, Transport Innovation and the Sustainable City of Tomorrow
Edited by William Riggs

Understanding Urban Cycling
Exploring the Relationship Between Mobility, Sustainability and Capital
Justin Spinney

Understanding Urban Cycling

Exploring the Relationship Between Mobility, Sustainability and Capital

Justin Spinney

Routledge
Taylor & Francis Group
LONDON AND NEW YORK

earthscan
from Routledge

First published 2021
by Routledge
2 Park Square, Milton Park, Abingdon, Oxon OX14 4RN

and by Routledge
605 Third Avenue, New York, NY 10017

First issued in paperback 2022

Routledge is an imprint of the Taylor & Francis Group, an informa business

Publisher's Note
The publisher has gone to great lengths to ensure the quality of this reprint but points out that some imperfections in the original copies may be apparent.

British Library Cataloguing-in-Publication Data
A catalogue record for this book is available from the British Library

Library of Congress Cataloging-in-Publication Data
Names: Spinney, Justin, 1973- author.
Title: Understanding urban cycling : exploring the relationship between mobility, sustainability and capital / Justin Spinney.
Description: New York : Routledge, 2020. | Series: Routledge equity, justice and sustainable city | Includes bibliographical references and index.
Identifiers: LCCN 2020021055 (print) | LCCN 2020021056 (ebook) | ISBN 9781138543225 (hardback) | ISBN 9781351007122 (ebook)
Subjects: LCSH: City planning–Environmental aspects. | Cycling. | Urban transportation.
Classification: LCC HT166 .S63755 2020 (print) | LCC HT166 (ebook) | DDC 796.6–dc23
LC record available at https://lccn.loc.gov/2020021055
LC ebook record available at https://lccn.loc.gov/2020021056

ISBN: 978-0-367-56773-6 (pbk)
ISBN: 978-1-138-54322-5 (hbk)
ISBN: 978-1-351-00712-2 (ebk)

DOI: 10.4324/9781351007122

Typeset in Goudy
by Wearset Ltd, Boldon, Tyne and Wear

Contents

Figures

Acknowledgements

There are a lot of people without whom this book would not have been possible. I'm not going to mention you all because it would take too long. Suffice it to say that without the conversations, feedback and encouragement of a lot of people over the last ten years, this book would either not be here or be a whole lot worse. So without trying to offend anyone by ranking this list, I'd like to thank:

Helen, Amy and Elliott, for reminding me that there's so more to life than work and bikes; Phil Crang, for starting it all off; Wen and Shuyi, for so many thoughtful collaborations; Tim Schwanen, for encouraging critiques; the late John Urry and the original Cycling and Societyers: John Parkin, Peter Cox, Tim Jones, Dave Horton, Kevin Kricek, Ben Fincham, Mark McGuinness, Paul Rosen; Sven, Malene, Katrine, and all at Cosmobilities; Rachel Aldred and Kat Jungnickel; Marco te Brömmelstroet, George Liu, all at the CRB; Mimi Sheller, Pennie Drinkall, Monica Buscher, Cosmin Popan and all at Lancaster; Kim Kullman, Pablo Bose, Denver Nixon, Natalie Koch, Tim Cresswell, Paula Jiron, Rob Imrie, Laura Golbuff, Lucy Baker, Richard Bower, Ben Spencer, Emma Street, Shaun Williams, Brian Deegan, Katrina Brown, Kiron Chaterjee, Juliet Jain, Billy Clayton, Graham Parkhurst, Rebecca Jones, Ben Coles, David Evans, Dimitris Potoglou, Rob Huggins, Oleg Golubchikov, Andrea Collins, Pete Adey, Pete Merriman, David Bissell, Tom Birtchnell, Alan Latham, Tim Jackson, Shane Fudge, Kate Burningham, Geoff Cooper, Nicky Green, Anna Nikolaeva, Graeme Sheriff, Mags Adams, Phil Pinch, Suzy Reimer; and apologies to anyone I have missed.

1 Cycling towards sustainability?

The cycling renaissance

Take a look around and you will see that cycling is making a comeback in many cities across the world, or is at the least represented as being a key feature of urban life that we should all embrace in order to be healthy, efficient and environmentally friendly. Some indexes now rank cities according to their cycle friendliness in relation to a host of factors, including bicycle usage, infrastructure, theft, and presence of bike-sharing, amongst others (Global Bicycle Cities Index 2019); and the ability to cycle safely in a city now forms a key pillar of many liveable city indexes.

Statistics give some sense of the scale of the renaissance: between 2010 and 2015, New York saw an 80% growth in cycling (Sutton 2016). In 2018, London saw its biggest ever increase in cycling, growing by 5% to more than 4 million km cycled (TfL 2019: 1). The GLA's current stated goal is to increase cycling levels to 5% mode share by 2025, which would mean an overall 400% increase (GLA 2014). A study of 12 neighbourhoods in Shanghai, which has until recently been quite hostile to cycling, saw a growth from 33.3% mode share to 48.3% mode share between 2017 and 2018 (Yingnan *et al.* 2019: 1). Across the water, Taipei, which has long been known as a centre of bicycle manufacture, has seen cycle mode share climb to around 5% (Van Mead 2016), with the most frequently used bike share system in the world (Smith 2014). Whilst these cities have a long way to go before their cycle numbers are anywhere close to those recorded in cities like Amsterdam and Copenhagen, with cycling mode shares of 32% and 35% respectively (Mason *et al.* 2015), they are representative of the way in which cycling is back on the urban mobility agenda.

However, a closer look at these cities reveals some less desirable facts. In all cases, most cycle journeys are to get to and from work, with much lower numbers engaged in 'social cycling' for leisure, journeys to school and social reproduction. In cities like London, the demographic profile is narrow – the average cyclist is a white middle-aged man on a good income. In cities like Shanghai and Taipei the opposite problem is evident, with cycling's lowly status meaning that most cyclists are from low-income and student populations. As this suggests, cycling has not conquered urban environments in the ways many would have liked, and indeed it is largely encouraged as a commuter mode, and even then one with appeal only to a very limited demographic who undertake simpler journeys, and possess the freedom from caring responsibilities and the cultural status (or lack of it) necessary to engage in cycling.

As this suggests, cycling in its commonly promoted variant is perhaps not as sustainable as it would first seem. By focusing in the first instance on facilitating commuting cycling and carefully managing other forms of cycling, what we see emerging is a 'weaponised' version of cycling that is more concerned to produce better human capital and faster circulation times than enable human flourishing; to help brand place and nation rather than produce places that enhance wellbeing; and to enable the work journeys of a minority rather than enhance the wellbeing of a diversity of citizens.

Congruently, a central argument of this book is that current manifestations of urban cycling do little to question the dominant growth paradigm of production and consumption, rather they reinforce it. Dominated by 'a conventional view of transport planning … which regards speed, individualism, [and] private mobility … as a desirable future' (Barr 2018: 8), I argue that whilst contemporary cycling policy may facilitate some relative decoupling of economic productivity from emissions, in most (if not all) cities where it is present, it is largely being employed to maintain an unsustainable growth agenda. Barr (2018) has argued that:

> there is an implicit mainstream political narrative that more travel is economically beneficial … within this overall narrative, only certain kinds of behavioral change are deemed acceptable or desirable; major behavioral change is not desirable because it neither fits with the economic narrative nor the ideology of the small state and consumer choice.
>
> (Barr 2018: 4)

Within this narrative we can include cycling: more and faster cycling is deemed beneficial, but to go beyond this and promote other qualities of cycling like slowness, socialising, sensory richness, place-making and playfulness is not encouraged because it does not fit the economic growth narrative and indeed what cities 'should be for'. Indeed, cycling in its current formations supports rather than contests 'a political-economic model bent on endless accumulation, economic growth, consumption and waste on a finite planet' (Figueroa-Helland *et al*. 2016: 260).

Accordingly this book sets out to make the case that cycling in its current formations represents a kind of 'mobility fix' because it is promoted primarily to 'fix' matters of economic concern where other forms of fixing are deemed politically and economically untenable. These matters of concern include speeding up circulation time of capital and labour; enhancing human capital; enhancing industrial and place image; and enabling the production and capture of 'behavioural surplus' (Zuboff 2019) for the manufacture of new products. Focusing on London, Shanghai and Taipei, the different chapters demonstrate the ways in which cycling as a social practice becomes 'economised' (Çalışkan and Callon 2010) through the valorisation of labour power. In the process of producing forms of cycling that maximise labour power and behavioural surplus, potential multiplicities of cycling are narrowed with regard to the qualities of it that are deemed important and valuable, and the users who can participate in it.

Ultimately the book makes the argument that the current trajectory of cycling reflects a weak approach to green economy because of its exclusivity and instrumentalisation in service to an unbridled growth agenda. In bringing this critical stance to cycling research, the book does not argue that more cycling is not desirable, but for a broader view of cycling that embraces a full range of qualities that could not only transform how it is experienced, but help to transform the goals to which it is oriented away from economic growth and toward human flourishing, connection and wellbeing.

A number of the chapters in this book are adaptations and reworkings of previously published work. Chapter 3 reworks some material from Spinney *et al.* (2017) on design; Chapter 4 is a reworking of Spinney (2016) on the mobility fix; Chapter 5 is a reworking of Spinney (2010) on South Bank cycling; Chapter 7 is a reworking of Spinney and Lin (2019) on fixing cycling in Taiwan. All the ideas in these papers have been substantially reworked to emphasise the overarching narrative of this book related to mobility fixing and the narrowing of cycling as a form of practice. Indeed, this narrative has slowly emerged from my research and writing of these papers over the last ten years, and this book is an attempt to bring these thoughts together in one place in a more or less coherent way.

I would point out that there is no research design or methodology chapter in the book. I have published extensively on mobile methodologies (Brown and Spinney 2010; Spinney 2011; Spinney and Jungnickel 2019). These and other empirical papers are the best place to find out in more detail about the methods used to gather data for the chapters. By way of summary, they encompass fairly straightforward in-depth interviews; ethnography; mobile video ethnography; and archival/desk research. I would like to thank everyone who has agreed over the years to take part in the various studies that underpin the interpretations found in this book.

From hero to zero: a brief history of cycling

Whilst cycling is evidently making a comeback in many cities around the globe, as the term comeback suggests, its fortunes have varied over the years since its 'invention' around 1860. After a period of extensive and often over-imaginative invention, it was in the late 1860s that the bicycle as we know it really began to materialise in the form of the velocipede (Dodge 1996: 17–19).

So popular was this machine that cycling took off in England in the late 1860s and early 1870s in what was widely regarded as a 'craze'. The cycling explosion of the 1890s in Britain, for example, was:

> greeted with ecstatic pronouncements of not only the limitless physical improvement it promised the masses, but also the opportunities for spiritual and social growth it offered. Every societal end from strengthening of the family to the curing of alcoholism and opium addiction was credited to the wonderful exercise of wheeling.
>
> (Whorton 1982: 33)

However, even in its earliest incarnations, cycling was not evenly distributed across the population: it was classed, gendered, and racialised. Despite its purported medicinal properties, cycling in its earliest phase was profoundly gendered. McCrone notes that:

> during the late-Victorian period, complicated social factors involving class and gender shaped the concept of what were suitable and acceptable physical activities for boys and girls, men and women, rich and poor. In other words, the use of the physical body was very much constrained by the social body, and rested at least partially on the assumption that class and gender differences involving patterns of dominance and subordination should apply to physical education and sport.
>
> (McCrone 1991: 179)

Congruently:

> women on bicycles in the early 1890s represented a serious incongruence, placing the audience in the dilemma of wondering how to respond to apparently respectable women (i.e., respectably dressed) who conspicuously flaunted themselves for the sake of a questionable activity.
>
> (Simpson 2003: 4)

However, change was rapid, and by the 1890s a woman on a bicycle was more likely to be deemed respectable than not (Simpson 2003: 6). Despite this, as both Raszeja and Ritchie suggest, women cyclists found it almost impossible to establish themselves legitimately in competitive sport, remaining largely spectators (Raszeja 1992: 6; Ritchie 1999: 514). Even with widening participation by women throughout the late nineteenth and early twentieth century, it was frowned upon up until the 1930s for a woman to race seriously on a bicycle (Ritchie 1999: 514).

In its early days cycling was also a classed phenomenon, with certain variants deemed more acceptable than others. As a consequence of the Victorian desire for 'rational recreation' and 'muscular exertion' over the more traditional leisure pursuits of the working class, certain types of cycling were given prominence over others. Trick cycling, for example, whilst often in evidence as a 'sideshow' at sport/commercial cycling events, was often frowned upon by some commentators who saw its protagonists as mere 'scramblers' on bikes compared with the true 'muscular athleticism' displayed by real riders (Ritchie 1999: 501). Trick cycling it seems had too much in common with working-class leisure pursuits such as the circus and the music hall and was consequently not considered athletic enough to be a legitimate leisure practice. Whilst trick riding was fairly common in the late 1870s, it gradually died away and was replaced by indoor or track racing. The decline of this style of movement to the dominance of racing and touring appears to be no coincidence in the context of a moralistic middle-class Victorian agenda that favoured athletic pursuits.

After the cycle boom

By 1900 the bicycle craze in Britain was effectively over and events were no longer commonly reported in the newspapers and magazines. Prices plummeted and people for whom cycling had been previously exclusive began to be able to afford it. Middle- and upper-class interest in the machine faded rapidly despite a decline in prices, and it rapidly became the tool of the working classes (Petty 1996: 126). By the 1930s the bicycle was very much a poor man's transport (Oddy 2000: 174). Consequently it became a mundane activity, an article of utility, 'to get clerks and workmen to and from their business' (Petty 1996: 125). As well as journeys to and from work, the bicycle also *was* work for many people, from messenger boys to postmen (Burstall 2004: 8).

It was not just that the bicycle was no longer new that put it out of favour; it was the meteoric growth of the modern motor car in the UK from the early 1900s – and particularly after the First World War – that accelerated its decline (see, for example, Morton 1933). After its first licensing on UK roads, by 1920, there were 187,000 private cars and 88,000 public vehicles on the UK's roads (Fullerton 1975: 30). After 1920 the growth in vehicles accelerated and outstripped the capacity of the existing roads, reaching 1 million by 1930, and 13 million by 1972 (Fullerton 1975: 33). This growth was accompanied by a concomitant increase in road accidents and fatalities, and increased congestion in towns and cities.

The increasing complexity and dangers posed by automobility consequently brought a new direction to transport planning which largely failed to consider pedestrians and cyclists (Bendixson 1974: 53). The net result of pro-car policies has been that the existing infrastructure and environment is deemed unsuitable for cyclists, a point qualified by Whitelegg (1997), who states that, 'the urban system has moved so far in the direction of satisfying the needs of the person in the car that it acts as an efficient deterrent to those who struggle on as pedestrians or cyclists'

(Whitelegg 1997: 136). Consequently, as Horton notes, fear of cycling is now the primary deterrent to urban cycling (Horton 2007: 133).

The reduction in domestic cycling in the UK was also compounded by the loss of the cycling industry, which during the 1970s and 1980s moved to South East Asia (most notably to Taiwan and then China) (Rosen 2002). The loss of the domestic cycle industry meant there was less of a business case for government to support cycling as a mode of transport, because domestic industry had little to gain from trying to increase cycling levels.

As a result, and despite the various booms in leisure cycling and the inter-war growth in bicycle commuting amongst many of the working class (Pooley *et al.* 2005: 18), cycling for transport purposes has been in terminal decline in the post-war period and reached rock bottom in the mid to late 1990s. In 1952, 23% of all journeys were made by bicycle, but this had dropped to 9% just a decade later (DfT 2006: 14). By 1973 the number of miles cycled was only 17% of what it had been 20 years before (Plowden 1985: 35). Following a further gradual decline, by 1993, only 4% of all journeys were made by bicycle (DfT 2006: 14).

Cycling in China and Taiwan

Whilst many nations have followed a path to very low cycling levels that echoes that of the UK, the story has of course not been the same everywhere (see, for example, Norcliffe 2015). The story of cycling began later in China from around 1900, when the upper classes started to import the bicycle as a luxury good. Bicycle ownership was very low in these early years and grew slowly due to the high cost of bicycles, which limited ownership to the very wealthy (Zhang *et al.* 2014: 319). In line with many other countries, in its early phases cycling was very exclusive of the working classes.

This situation changed quite radically from 1978, when China instituted its great economic reformation and embarked on a more industrial future. In this period the bicycle became widely available to low-income families and cycling as a mode of everyday transport took off, with mode share in some cities approaching 50% (Zhang *et al.* 2014: 320). It is in this (relatively short) period between 1978 and 1995 that China was considered a cycling nation.

However, beginning in the 1990s, when many European and American cities were attempting to recuperate the bicycle, many Chinese cities were trying to phase it out because it was seen as inefficient and a symbol of backwardness. The decline in cycling in many cities was accelerated by government policy that actively sought to reduce cycling mode share. Central to this was the 1995 document 'Standard of Urban Road Traffic', which positioned the bicycle as a source of traffic conflict that should be reduced to enable increases in public transport and car use. Many city transport strategies took this on board: the 1993 Transport Plan for Guangzhou, for example, set a target of reducing cycling mode share from 33.8% in 1992 to 13.3% by 2010 (Zhang *et al.* 2014: 318).

Across the Strait in Taiwan, the story shares some similarities. Bicycles were introduced in Taiwan by the Japanese around 1903, with all having to be registered and taxed until 1973 (Chang *et al.* 2008: 32). Bicycle numbers grew steadily in cities like Taipei, reaching a peak of one bicycle per household in 1960 and a mode share of around 19% (Chang *et al.* 2008: 32–34). However, numbers declined from this peak, reaching a low of around 0.2 bikes per household in 1990 as city governments focused more on public transportation and the average income increased rapidly.

Ironically it was in no small part due to its success as a manufacturer of bicycles from the 1970s onwards that saw Taiwanese incomes rise, leading to a corresponding drop in the levels of bicycle use for daily mobility (Chang *et al.* 2008: 34). Whilst there was a nascent cycle industry from around 1945, fuelled by a large number of SMEs quick to adapt to bicycle component manufacture (Chu 1997: 55), from the 1970s Taiwan became a centre of manufacturing excellence as bicycle manufacture disappeared from European and North American nations and moved eastward (Rosen 2002; Norcliffe 2015). As a result of its newly industrialising status and the wealth that this brought, the car and moped became staples of Taiwan's urban transport a couple of decades before China. Writing in 2008, Chang *et al.* note that with a population of 23 million, Taiwan had 5.7 million cars, 12 million mopeds but only 1 million bicycles. From the 1990s, where cycling did feature in government policy, the focus was mainly on recreational cycling (Chang *et al.* 2008: 30, 36) and mode share in the capital city of Taipei was around 1% by 2002 (ibid.).

However, the downturn in the fortunes of the bike in Taiwan were also due to an influx of mopeds from 1968 onwards. In this year, a huge number of 49cc mopeds that did not require licensing to ride were imported from Japan and these proved more favourable as a means of daily mobility (Chang *et al.* 2008: 38). This also impacted the domestic industry, which suffered greatly until the US company Schwinn placed a 1 million unit order in 1972 (ibid.). Since then, the cycle industry has experienced many highs and lows, as I discuss in Chapter 7, but it remains important to the Taiwanese economy, with 95% of bicycles being manufactured for export (Chang *et al.* 2008: 39).

Every dog has its day

As these brief overviews attest, cycling has very diverse economic, social and cultural geographies that have been informed by economic globalisation, colonisation, gendered and classed differences, technological innovation, and socio-legal shifts, amongst others. Whilst they all suggest that different nations have arrived at low points in cycling levels at slightly different times, they all have in common the fact that for business and government, the bicycle's centrality to processes of capital accumulation – whether through the sale of bikes as a consumer object, creation of new markets, getting goods to market, or workers to work – is effectively deemed to be over by the 1990s. In these accounts, the bicycle is superseded by cars or mopeds on a developmental mobility continuum, where the car represents the ultimate evolutionary end state. As Zuev *et al.* (2018) argue, just as in the UK, nations like China have debated which mobility technologies are more civilising, with a conclusion that 'tends to favour the car as it is seen as a "next stage" (of development) after the bicycle' (Zuev *et al.* 2018: 8). By the 1980s and 1990s the bike is regarded as old technology: a transition mobility that has had its day and will slowly fade into the background as a niche past-time before fading completely from memory. The new policy directives and corresponding spatial fixes (in the form of new and bigger roads) implemented in order to accommodate the car helped to ensure that the bike would remain marginal because it could not (or was not allowed to) safely or efficiently operate within environments designed with the affordances of motor vehicles in mind.

Yet something unexpected has happened to bring cycling back from the dead. Despite its evident humanity (Miller 2001; Sheller 2004; Ferguson 2009), the car's place in the city has been increasingly problematised due to its capacity to generally diminish the liveability of place (Sennett 1994; Litman 2011). The normative assumption that the car is the future and the pinnacle of mobility development is being questioned. In response to a by now familiar set of matters of concern, cycling is positioned as a panacea for city economies slowed by cars and lack of innovation; unhealthy sedentary workers; polluted local and global environments; and cities whose image needs to be more liveable. As Stehlin (2014) has commented, many advocates point to 'its importance in contributing to urban vitality, freedom, invigorated commercial districts and sociability' (22). Accordingly, since the early 2000s, Shanghai, Taipei and London (in line with many other cities) have all adopted pro-cycling policies.

Since the early 2000s, Taiwan has adopted a more pro-cycling strategy with a commitment to build a 900 km cycle network around the island, and more limited development of cycle lanes in cities like Taipei. In 2008 central government officially committed to the development of a low-carbon society within which the bike is seen as an important component (Chang *et al.* 2008: 46). Whilst cycling levels remain low, Taipei has been successful with its 'You Bike' sharing scheme set up in 2012, which has helped boost the profile of the Bicycle Kingdom as more than just a centre of manufacture (Chen and Lu 2015: 1104). However, cycling for daily mobility in Taiwan remains the preserve of the young and those on low incomes. Some 87% of You Bike users, for example, are aged under 44 with 60% earning less than US$1000 per month and 70% educated to degree level or above (Chen and Lu 2015:1110) – effectively a student/low-income demographic.

In the current climate, the stance toward cycling in China varies from city to city, but many, including Beijing, Shanghai, Dalian and Guangzhou, have adopted pro-cycling transport strategies, including targets to increase cycling, building new separated cycle lanes, enabling public bike sharing systems and repealing bans on electric bicycles (Zhang *et al.* 2014: 318). More recently, as I discuss in more detail in Chapter 6, a new wave of dockless public bike-sharing has swept through China (and many other nations), further boosting the profile of cycling in its cities. However, the prevalence of cycling varies widely from city to city, with Zhang *et al.* (2014) reporting that whilst cities like Tianjin had close to 50% mode share in 2003, Chongqing had a mode share of around 1% in 2002 (323). Thus the contemporary mobility landscape in Chinese cities is varied, but also more complex than Western counterparts, with a much greater proportion of powered two-wheelers (PTW) used for daily mobility needs (Zuev *et al.* 2018), a situation that further complicates the re-entry of cycling into these urban transport systems. Whatever the case, the bicycle in the Chinese context is widely regarded as something used by those who cannot afford the next step up, such as the moped or e-bike (Norcliffe 2015: 15).

In the UK, since the 1990s there has been an increasing emphasis on reducing the harmful environmental effects of mobility. Consequently, efforts have been made in recent years to increase the numbers of people cycling, with a particular emphasis in London. These efforts have primarily been aimed at increasing access to cycling as a transport practice and have taken the form of either altering infrastructure to facilitate cycling (with mixed results), or 'soft' marketing campaigns, and training and advice aimed at giving people the skills and confidence required to cycle. In line with many other cities, Transport for London (TfL) also launched a public bike-sharing scheme (PBSS) on 30 July 2010. Initially confined to a 45 km^2 area of central London with 315 docking stations and 5,000 bikes, this was extended east in March 2012 to encompass a total of 575 docking stations and 8,000 bikes (Goodman and Cheshire 2014: 273). The current rate of usage is roughly 30,000 hires per day.

Yet this revalorisation of cycling has a very uneven texture: whilst the initial brief history in this introduction illustrated that cycling has always at different times and in different places been an exclusive practice along lines of gender, income, class and race, in its current iterations (cycling even where it has become popular and trendy) attracts only a narrow group of users that do not represent the wider city demographic. In London and the UK, the average cyclist is a middle-aged white man on a high income; in Shanghai, the average cyclist is low-income and/or a student; and in Taipei, they are a highly educated young person. In all cases they are able-bodied and less likely to have any caring responsibilities or impairments.

The spatiality of cycling and the purposes to which it is oriented are also uneven. It is notable that any inducement to cycle is overwhelmingly related to travel to places of work or (higher) education. London's cycle superhighway network enables quicker access to central London; Taipei's U Bike is focused around Metro stations to facilitate the last mile; and Shanghai's dockless bike-sharing companies have focused on specific public transport and employment sites where revenues can be maximised. Such a selective geography suggests that cycling has been brought back into cities because of its potential to solve matters of economic concern: here cycling is a tool that enables traffic to flow more efficiently with fewer delays and increased street capacity; to link up currently disconnected public transport modes for 'the last mile'; a form of movement that enables workers to be healthier and more productive; a new vehicle through which the internet of things can smarten the city and enable it to function more effectively; a way in which retail-led developments can be 'spectacularised'; and a new way in which cities can advertise themselves as more progressive and liveable in the era of global competition. In short, cycling is being imagined as a kind of 'fix' (Harvey 1982, 1991; Spinney 2016) to a host of urban problems that means only certain qualities of it are deemed to be of value and therefore encouraged. In line with this problematisation, a central argument of this book is that contemporary manifestations of urban cycling represent a 'mobility fix' (Minn 2013) because they represent a way of governing and fixing problems of accumulation and competitiveness that works primarily through producing alternative ways of moving rather than producing space (Baerenholdt 2013).

Cycling, mobility and productivity

Appleyard *et al.* (1964) observed the distancing and isolating experience of the car as 'a sequence played to the eyes of a captive, somewhat fearful, but partially inattentive audience, whose vision is filtered and directed forward' (cited in Venturi *et al.* 1972). Venturi *et al.*, in their landmark work of architectural critique, noted how such distance, speed and isolation had produced new landscapes, where sign becomes more important than form, and different users are separated for safety and efficiency. Arguably current manifestations of cycle design do little to move away from such imaginaries, where the aspiration remains moving the highest number of people through a given space as fast as possible. Certainly 'most governments have argued that economic growth is linked to higher traffic levels and that to seek to suppress road building equates to an attack on the principle of economic growth' (Marsden *et al.* 2014, cited in Barr 2018: 3). The same argument seems to be applied to the promotion of cycling: cycling enables public health to be improved, productivity to grow, and city image to be enhanced, so why would we stop? The underlying argument remains the same: more mobility means greater growth. There is no thought for the character of that mobility or any critical perspective on the goals (of unrestrained production and consumption) it is being oriented towards achieving. As Barr has argued, 'associating emissions with transport in general hides the wider impact of transport as a producer of certain kinds of economic activity' (Barr 2018: 2). As this suggests, whilst lowering emissions and decoupling (Jackson 2009) have dominated the debate, it also matters what forms of social and economic activity transport is oriented toward. As Jackson points out, 'over the last 30 years, what little relative decoupling we have seen has been offset by economic growth' (2009: 8).

This dilemma would appear to have been ignored in the case of current cycling policy. In its current variants, cycling is stripped of alternative qualities to intensify a focus only on those that can enhance productivity, whether that be through speed, spectacle, surplus or image. Congruently, what we seem to be producing is a cycling system that attempts to mimic the car system, just slightly slower (or indeed, in the context of current high levels of congestion, slightly faster). Contemporary manifestations of cycling are thus economised in service to a 'green growth' agenda, where maximal economic growth is pursued but with 'significant environmental protection' (Ferguson 2015: 19).

Ultimately, the cycling-related solutions adopted to address different 'matters of concern', whether in relation to under-performing industry, road spaces, retail spaces or city image, are not just technical in nature: what we see emerging are profoundly cultural measures spread over disparate actors, materials, discourses, and techniques that all revolve around rendering the social economic in various ways. As a result, this book is not only concerned to offer an explanation for why we see cycling manifest in particular ways, but also to try and show how cycling as a social practice is rendered economic, or as Çalışkan and Callon (2009) define it, 'economised'. Here the term economisation 'is used to denote the processes that constitute the behaviours, organizations, institutions and, more generally, the objects in a particular society which are tentatively and often controversially qualified, by scholars and/or lay people, as "economic"' (Çalışkan and Callon 2009: 370).

The individual chapters demonstrate that mobility fixing works through processes of economisation. Different chapters in the book show how modern planning and bicycle design narrow the range of users and uses that can participate in the cycling boom (Chapter 3); how cycling is promoted as a way to speed up the circulation time of capital and improve the quality of human capital (Chapter 4); how playful forms of cycling are managed as a source of behavioural surplus that can enhance the dynamism of waterfront retail (Chapter 5); how cycling is used to enhance the competitiveness of domestic industry, the city and nation through adopting a cycling culture (Chapter 6); and how cycling is used as a vehicle to facilitate production and collection of valuable user data in the case of PBSS (Chapter 7). All of them revolve around harnessing and shaping the qualities of cycling as a social practice to benefit economic ends. Here particular variants of cycling are brought into being through often gradual agencements of landscapes, actors, competencies, regulations, management strategies, designs and discourses, etc., that shape particular terrains and corresponding mobile behaviours in order to render some forms of cycling more likely to emerge than others (Krivy 2018: 15).

Accordingly, central to understanding processes of economisation are concepts of labour power and behavioural surplus. It is through these that the labour of cycling is valorised in relation to particular matters of concern and manufacturing processes. The book draws upon insights from governmentality and political economy to argue that processes of economisation are principally bio-political in that they seek to operate through shaping individuals to act upon themselves within particular framings. In this reading, economisation is made possible because particular appropriate forms of movement are fostered through the construction of particular discourses, technologies and places that enable their performance whilst discouraging others (Galloway, in Krivy 2018: 19).

Boltanski and Thevenot argue that capitalism has become increasingly predatory, preying on the 'people, ideas, things and movements that are in direct opposition to it' (cited in Mould 2018: 13). Capitalism, as Mould (2018) has argued, has become extremely effective at co-opting its own critics and in so doing redefining and producing a variant of practice (such as cycling) best suited to feed its own growth. In line with this, Mould argues in the case of promoting creativity in cities that capitalism has become increasingly effective at stabilising and defusing such practices by 'naming them, offering them a place at the table, mainstreaming them and bringing them into the public consciousness, prepping them for commercialisation' (ibid.). One result of this is that other readings of such practices that could help to create alternative social worlds are filtered out if they do not fit the need for profit.

Yet there is nothing inevitable about the particular versions of cycling we see being managed back into cities. It is easy to believe that these are the only variants of cycling that can be brought back into cities, or that these are the only reasons why cycling should be brought back into cities, yet cycling practices are diverse: cycling, for example, is understood as a plural practice that has embraced multiple qualities: lauded for its playful and sensory qualities (Clayton *et al.* 2017; Spinney 2006, 2010; Stevens 2007); been adopted as a vehicle for emancipation (Jungnickel 2015); political protest (Aldred 2012, Cox 2017, Horton 2006); and alternative lifestyle (Fincham 2006, Furness 2005); and been an inclusive practice enjoyed by different age groups (Jones *et al.* 2016).

Accordingly we could imagine a situation where the slowness and openness of cycling produce new spaces, cultures and identities that emphasise alternative qualities and are not geared towards facilitating it as purely transport, or in facilitating the journey to work. If the priorities for those governing cities were less concerned with harnessing the economic potential of social practice, and more concerned with broader ideas of human wellbeing and flourishing, arguably the variants of cycling we witness in cities could be more varied: PBSS could be found in areas that are predominantly residential, could include more than just standard bicycles, and could be transparent in their use of data; cycle superhighways need not be bicycle motorways designed to maximise speed and minimise interaction, but rather designed and located in ways that lead to rises in social connection, greater user diversity, place attachment, and increased enjoyment; playful forms of cycling could be accommodated in multiple places and used to encourage greater diversity of participation rather than ghettoised in service to 'culture'. Such a cycling system moves away from a principle where growth (and hence the individual commuter) is prioritised, and towards one where – decoupled from economic growth – a broader understanding of happiness and wellbeing for all is prioritised.

I can well imagine (and indeed your voices have often been in my head whilst writing this book) that many cycle activists and scholars will be questioning such a critical stance. Surely the fact that cycling is back on the menu is something to be celebrated? And of course it should be; it has required a lot of time and energy to recuperate cycling to the point it is at now and it should be applauded. I am also a cycling advocate and this is not a call to halt the cycling agenda because of its flaws; rather it is a call to rescue the cycling agenda from technocrats and economists and a potentially delusional decoupling fantasy (Fletcher and Rammelt 2017): through critical analysis the aim should be to ensure that cycling achieves its potential as a vehicle for reshaping our emotional and physical associations with place, rather than a narrowly framed one of improving productivity, physical health or city image. The production of commuting cycling subjects is not a wholly negative development and can be seen as a necessary step toward a more sustainable system. The danger is that it becomes the end goal and indeed substitute for a sorely needed and more radical and post-growth reimagining of what cities might do.

The book's central message therefore is that the manner in which cycling is being popularised – bringing cycling into cities through entrepreneurial circuits of capital and to solve economic crises – means that the versions of cycling we are witnessing are somewhat desiccated ones that reflect and reproduce a narrow vision of the city as primarily geared towards economic growth rather than human flourishing. Certainly it is increasingly likely that societal and environmental needs are not being met by a logic of cycle promotion geared toward sustaining capital investment and industrial growth; and distributed upon a consumer logic to those of higher socio-economic status rather than one of meeting transport need and increasing accessibility.

In line with this, the book seeks to highlight the importance of academic thought in (re)framing the possibilities for cycling. As Çalışkan and Callon have argued in relation to academic economics, such work is performative because of the ways in which ideas around what it is and could be are increasingly indivisible from its agencement in spheres of practice and policy-making (Çalışkan and Callon 2010: 19). Likewise, the performative function of whatever versions of 'cycling' are described and deployed by social scientists help to frame cycling policy and practice. As a result, we need as broad a debate as possible regarding what could constitute cycling and what it might be for. Congruently, we need to agitate for variants of cycling that are not enslaved to economic growth, and where success is measured through a more holistic idea of wellbeing, where choice, fun, diversity, inclusivity and participation are foregrounded. As Wetzel and Wetzel argued some time ago, the obstacles to meaningful change remain enormous if GNP continues to be the key measure of wellbeing (Wetzel and Wetzel 1995: 18).

References

Aldred, R. (2012) Governing transport from welfare state to hollow state: The case of cycling in the UK, *Transport Policy*, 23, pp. 95–102.

Appleyard, D., Lynch, K. and Myer, J.R. (1964) 'The view from the road', quoted in Venturi, R., Scott Brown, D. and Izenour, S. (eds) *Learning from Las Vegas: The forgotten symbolism of architectural form* (Cambridge, MA: MIT Press, 1972), 74.

Baerenholdt, J.O. (2013) Governmobility: The powers of mobility, *Mobilities*, 8(1), pp. 20–34.

Barr, S. (2018) Personal mobility and climate change, *Advanced Review*, pp. 1–19.

Bendixson, T. (1974) *Instead of Cars* (London: Temple Smith).

Brown, K. and Spinney, J. (2010) 'Catching a glimpse: the value of video in evoking, understanding and representing the practice of cycling', in Fincham, B., McGuinness, M. and Murray, L. (eds) *Mobile Methodologies* (Basingstoke: Palgrave-Macmillan), pp. 130–151.

Burstall, P. (2004) *The Golden Age of the Bicycle: The worldwide story of cycling in the 1890s* (Marlow on Thames: Little Croft Press).

Çalışkan, K. and Callon, M. (2009) Economization, part 1: Shifting attention from the economy towards processes of economization, *Economy and Society*, 38(3), pp. 369–398.

Çalışkan, K. and Callon, M. (2010) Economization, part 2: A research programme for the study of markets, *Economy and Society*, 39(1), pp. 1–32.

Chang, J., Chang, H.W. and Lee, Y.K. (2008) 'The status of cycling in Taiwan', in Tiwari, G., Arora, A. and Jain, H. (eds) *Bicycling in Asia* (Netherlands: Interface for Cycling Expertise, I-CE).

Chen, S-Y. and Lu, C-C. (2015) A model of green acceptance and intentions to use bike-sharing: You Bike users in Taiwan, *Network Spatial Economics*, 16, pp. 1103–1124.

Chu, W.W. (1997) Causes of growth: A study of Taiwan's bicycle industry, *Cambridge Journal of Economics*, 21, pp. 55–72.

Clayton, W., Parkin, J. and Billington, C. (2017) Cycling and disability: A call for further research, *Journal of Transport and Health*, 6, pp. 452–462.

Cox, P. (2017) 'Bicycle design history and systems of mobility', in Spinney, J., Reimer, S. and Pinch, P. (eds) *Mobilising Design, Designing Mobilities: Intersections, affordances, relations* (Abingdon: Routledge), pp. 49–61.

Department for Transport (2006) *Transport Statistics Great Britain* (London: The Stationery Office).

Dodge, P. (1996) *The Bicycle* (Paris-New York: Flammarion).

Ferguson, H. (2009) Driven to care: The car, automobility and social work, *Mobilities*, 4(2), pp. 275–293.
Ferguson, P. (2015) The green economy agenda: Business as usual or transformational discourse?, *Environmental Politics*, 24(1), pp. 17–37.
Figueroa-Helland, L., Lindgren, T. and Pfaeffle, T. (2016) Civilization on a crash course? Imperialism, subimperialism and the political-ecological breaking point of the modern/colonial world-system, *Perspectives on Global Development and Technology*, 14, pp. 257–289.
Fincham, B. (2006) Bicycle messengers and the road to freedom, *Sociological Review*, 54(s1), pp. 208–222.
Fletcher, R. and Rammelt, C. (2017) Decoupling: a key fantasy of the post-2015 sustainable development agenda, *Globalizations*, 14(3), pp. 450–467.
Fullerton, B. (1975) *The Development of British Transport Networks* (Oxford: Oxford University Press).
Furness, Z. (2005) *Put the Fun between Your Legs! The politics and counterculture of the bicycle*, unpublished PhD thesis (University of Pittsburgh).
Global Bicycle Cities Index (2019) Available at: www.coya.com/bike/index-2019 [accessed January 2020].
Goodman, A. and Cheshire, J. (2014) Inequalities in the London bicycle sharing system revisited: Impacts of extending the scheme to poorer areas but then doubling prices, *Journal of Transport Geography*, 41, pp. 272–279.
Greater London Authority (2014) *Cycle Flows on the TFL Road Network* (London: GLA). Available at: http://data.london.gov.uk/dataset/cycle-flows-tfl-road-network [accessed 26 November 2014].
Harvey, D. (1982) *The Limits to Capital*, 2nd edition (London: Verso).
Harvey, D. (1991) *The Condition of Post-Modernity: An enquiry into the origins of cultural change* (Malden, MA: Wiley- Blackwell).
Horton, D. (2006) Environmentalism and the bicycle, *Environmental Politics*, 15(1), pp. 41–58.
Horton, D. (2007) 'Fear of cycling', in Horton, D., Rosen, P. and Cox, P. (eds) *Cycling and Society* (Aldershot, UK/Burlington, VT: Ashgate), pp. 133–152.
Jackson, T. (2009) *Prosperity without Growth: The transition to a sustainable economy* (London: Sustainable Development Commission).
Jones, T., Chatterjee, K., Spinney, J., Street, E., Van Reekum, C., Spencer, B., Jones, H., Leyland, L.A., Mann, C., Williams, S. and Beale, N. (2016) *Cycle BOOM: Design for lifelong health and wellbeing: Summary of key findings and recommendations* (Oxford: Oxford Brookes University).
Jungnickel, K. (2015) 'One needs to be very brave to stand all that': Cycling, rational dress and the struggle for citizenship in late nineteenth century Britain, *Geoforum*, 64, pp. 362–371.
Krivy, M. (2018) Towards a critique of cybernetic urbanism: The smart city and the society of control, *Planning Theory*, 17(1), pp. 8–30.
Litman, T. (2011) *Developing Indicators for Comprehensive and Sustainable Transport Planning* (Victoria, Canada: Victoria Transport Policy Institute).
Mason, J., Fulton, L. and McDonald, Z. (2015) *A Global High Shift Cycling Scenario* (Davis, CA: Institute for Transportation and Development Policy/University of California, Davis).
McCrone, K. (1991) Class, gender, and English women's sport, c. 1890–1914, *Journal of Sport History*, 18(1), pp. 159–182.

Miller, D. (ed.) (2001) *Car Cultures* (Oxford: Berg).

Minn, M. (2013) The political economy of high speed rail in the United States, *Mobilities*, 8(2), pp. 185–200.

Morton, H.V. (1933) *The Pageant of the Century* (London: Odhams Press).

Mould, O. (2018) *Against Creativity* (London: Verso).

Norcliffe, N. (2015) *Critical Geographies of Cycling: History, political economy and culture* (Farnham: Ashgate).

Oddy, N. (2000) 'Cycling in the drawing room', in Ritchie, A. and van der Plas, R. (eds) *Proceedings, 11th International Cycle History Conference* (San Francisco), pp. 169–176.

Petty, R. (1996) 'Women and the wheel: The bicycle's impact on women', in van der Plas, R. (ed.) *Proceedings, 7th International Cycle History Conference* (Anaheim, CA: KNI), pp. 112–131.

Plowden, S. (1985) *Transport Reform: Changing the rules* (London: Policy Studies Institute).

Pooley, C., Turnbull, J. and Adams, M. (2005) *A Mobile Century? Changes in everyday mobility in britain in the twentieth century* (Aldershot: Ashgate).

Raszeja, V. (1992) A decent and proper exertion: The rise of women's competitive swimming in Sydney to 1912, *ASSH Studies in Sport's History*, No. 9.

Ritchie, A. (1999) The origins of bicycle racing in England: Technology, entertainment, sponsorship and advertising in the early history of the sport, *Journal of Sport History*, 26(3), pp. 489–520.

Rosen, P. (2002) *Framing Production: Technology, culture and change in the British bicycle industry* (Cambridge, MA: MIT Press).

Sennett, R. (1994) *Flesh and Stone* (London: Faber & Faber).

Sheller, M. (2004) Automotive emotions: Feeling the car, *Theory, Culture and Society*, 21(4–5), pp. 221–242.

Simpson, C. (2003) Managing public impressions: Strategies of nineteenth century female cyclists, *Sporting Traditions*, 19(2), pp. 1–15.

Smith, G. (2014) *Pedaling into the Future*. Available at: https://taiwantoday.tw/news.php?post=14178&unit=8%2C8%2C29%2C29%2C32%2C32%2C45 [accessed January 2020].

Spinney, J. (2006) A place of sense: A kinaesthetic ethnography of cyclists on Mt Ventoux, *Environment and Planning D: Society and Space*, 24(5), pp. 709–732.

Spinney, J. (2010) 'Mobilising sustainability: Partnership working between a pro-cycling NGO and local government in London (UK)', in Fudge, S., Peters, M. and Jackson, T. (eds) *Low Carbon Communities: Imaginative approaches to combating climate change* (Cheltenham, UK: Edward Elgar), pp. 89–107.

Spinney, (2011) A chance to catch a breath: Using mobile video ethnography in cycling research, *Mobilities*, 6(2), pp. 161–182.

Spinney, J. (2016) Fixing mobility in the neoliberal city: Cycling policy and practice in London as a mode of political–economic and biopolitical governance, *Annals of the American Association of Geographers*, 106(2), pp. 450–458.

Spinney, J. and Jungnickel, K. (2019) 'Studying mobilities', in Atkinson, P., Delamont, S., Cernat, A., Sakshaug, J.W. and Williams, J.W. (eds) *Sage Research Methods Foundations* (London: Sage Publications).

Spinney, J. and Lin, W. (2019) (Mobility) fixing the Taiwanese bicycle industry: The production and economisation of cycling culture in pursuit of accumulation, *Mobilities*, 14(4), pp. 524–544.

Spinney, J., Reimer S. and Pinch, P. (eds) (2017) *Mobilising Design* (Oxford: Routledge).

Stehlin, J. (2014) Regulating inclusion: Spatial form, social process, and the normalization of cycling practice in the USA, *Mobilities*, 9(1), pp. 21–41.
Stevens, Q. (2007) *The Ludic City: Exploring the potential of public spaces* (Oxford: Routledge).
Sutton, M. (2016) *33 Key Cities Where Cycling is Growing Its Modal Share*. Available at: https://cyclingindustry.news/five-key-cities-where-cycling-is-taking-modal-share-from-cars/ [accessed January 2020].
Transport for London (2019) *Cycling Trends Update*. Available at: http://content.tfl.gov.uk/cycling-trends-update.pdf [accessed 26 November 2019].
Van Mead, N. (2016) Return of the bicycle kingdom? How pavement cycling is transforming Taipei, *The Guardian*, 15 March. Available at: www.theguardian.com/cities/2016/mar/15/bicycle-kingdom-reborn-pavement-cycling-taipei-taiwan [accessed 5 July 2020].
Wetzel, K. and Wetzel, J. (1995) Sizing the earth: Recognition of economic carrying capacity, *Ecological Economics*, 12, pp. 13–21.
Whitelegg, J. (1997) *Critical Mass: Transport, environment and society in the 21st century* (London: Pluto Press).
Whorton, J. (1982) 'Athlete's heart': The medical debate over athleticism, 1870–1920, *Journal of Sport History*, 9(1), pp. 30–52.
Yingnan, J., Ding, D., Gebel, K., Chen, L., Zhang, S., Ma, Z. and Fu, H. (2019) Effects of new dock-less bicycle-sharing programs on cycling: A retrospective study in Shanghai, *British Medical Journal Open*, doi:10.1136/bmjopen-2018-024280.
Zhang, H., Shaheen, S. and Chen, X. (2014) Bicycle evolution in China: From the 1900s to the present, *International Journal of Sustainable Transportation*, 8(5), pp. 317–335.
Zuboff, S. (2019) *Surveillance Capitalism: The fight for a human future at the new frontier of power* (New York: Hachette Book Group).
Zuev, D., Tyfield, D. and Urry, J. (2018) Where is the politics? E-bike mobility in urban China and civilizational government, *Environmental Innovations and Societal Transition*, doi:10.1016/j.eist.2018.07.002.

2 Towards a political-economy of cycling

Introduction

The main purpose of this book is an attempt to explain contemporary manifestations of urban cycling: why do we see particular variants of cycling (such as commuting) taking priority? Why do we see particular user groups dominating or marginalised? How do social practices of cycling become enrolled in ostensibly economic endeavours? In order to set up the discussions that follow, the chapter first reviews two overlapping sets of literature to understand why these questions are important, and the extent to which they have been engaged with in previous research: first, it summarises the transport literature as assuming a narrow 'productivist' framing of cycling in the shape of commuter-focused infrastructure and promotion. It argues that this literature has until recently been uncritical of the exclusive nature of much bicycle promotion with little emphasis on equity and justice; and still ignores the incorporation of cycling into global circuits of capital as a process requiring critical engagement. Indeed, in such accounts, the qualities that are valued and deemed important in cycling (zero emissions, healthy, space-efficient) are seen to be self-evident rather than constructed by particular actors (including academics). Second, the chapter reviews the mobilities literature in relation to cycling. It argues that whilst this literature has provided new theoretical and critical insights around embodiment, diversity and governmentality in relation to cycling, it has until very recently marginalised the relationship between cycling and broader political-economic formations. Building upon this, the second section brings together work in political economy, governmentality and economic sociology to provide a framework for understanding why mobility matters to urban economies, and more specifically, through processes of economisation and agencement, *how* ostensibly social mobility practices become instrumentalised and enrolled in circuits of capital accumulation.

Cycling in transport scholarship

Cycling scholarship through the 1990s and into the early 2000s pushes off from (and in truth very much constructs) a growing awareness of the harms wrought by automobility, particularly rising levels of congestion, pollution and road death (Horton 2006: 43, see also Horton 2007; Vigar 2002). Against a backdrop of very low levels of cycling in almost all Western nations, it is very much characterised by a pragmatism borne of cycling's lowly status. Scholarship in this period is concerned with mapping particular historical trends within both sporting and utility cycling, and documenting ways in which cycling can be 'mainstreamed', most notably through an emphasis on spatial planning, design and policy revision. As a consequence, scant attention has been paid to issues of justice, diversity and equity within this body of research, preferring instead to construct the cyclist as a somewhat homogenous user.

Earlier research on cycling positions it within the field of transport geography as a panacea to the ills created by automobility. Whilst it is impossible to characterise the substantial body of cycling research as any one thing, and there is evidence that a more nuanced approach is emerging (see, for example, the travel diary related work of Axhausen *et al.* 2000), it is fair to say that much of it has been preoccupied with the health and environmental benefits associated with cycling; the material and 'rational' factors that influence why people do or do not cycle; and the dangers of cycling. Certainly the majority of earlier research on cycling has situated it under the heading of transport and brought with it a largely quantitative methodological toolkit (see, for example, Aultman-Hall *et al.* 1997; Cervero and Radisch 1996; Crane and Crepeau 1998; McClintock 1992; Parkin 2003; Parkin *et al.* 2007) geared towards understanding cycling (and indeed movement in general) as a rational process predicated on decision-making and utility maximisation whose chief determinants are cost, time, distance and the quality of infrastructure and facilities.

Providing an evidence base for the pro-environmental credentials of cycling has been a key theme mobilised in attempts to recuperate it as a valid mode of transport (Cupples and Ridley 2008). In a study of transport in university campuses, Tolley (1996), for example, focuses on the extent of environmental benefits that could accrue from a shift to cycling amongst staff and students. Lumsdon and Tolley (2001) conclude that:

> although bicycles require some resources in manufacture, they are benign in use ... When combined with its efficient use of space, low energy consumption, speed over short distances, reliability and affordability, the bicycle is a sustainable means of access for most sectors of the population.
>
> (Lumsdon and Tolley 2001: 293)

Similar accounts of the benefits of cycling in contrast to those of the automobile can be found in Barr (2003), Bickerstaff and Walker (1999, 2003), Button and Nijkamp (1997), Cohen (2006), Greene and Wegener (1997), Kearns *et al.* (2003), Lattimer and Munro (2006), Lucas (1998), and Newman *et al.* (1995).

Paralleling this has been an emphasis on gathering evidence for other pro-social aspects of cycling in contrast to the car, particularly in relation to traffic accidents and public health outcomes. Like many earlier papers related to cycling, Whitelegg (1983) reports on the role of the motor car and a related landscape of engineering and lack of alternatives that contributes to a high road traffic accident casualty rate. In line with many papers advocating cycling, Whitelegg points to the poor road safety record of the car to implicitly make a case for increasing cycling and public transport. Writing in 1994, Tolley discusses the need for improved accessibility and better health outcomes from transport, emphasising, for example, cycling's health benefits for employers who encourage staff to shift to cycling (70). A number of other authors have also lamented the destructive capacities of driving at the scale of the fleshy body (cf. Adams 2001; Bohm *et al.* 2006; Bonham 2006; Dery 2006; Fincham 2006; Freund and Martin 2001; Redshaw 2007; Whitelegg 1987).

Whilst far from writing off other meanings of cycling, many such accounts say little about other practices of cycling, suggesting rather that the primary logic to promote cycling is to replace the car within the context of auto-dominated transportation systems. As Cupples and Ridley (2008) have argued, 'cycling is thus mobilised as a way in which an individual can meet the "civic obligation to moderate the burden of risk which he or she imposes on society"' (256). Implicit within this, the sustainability of economic growth as a goal remains unquestioned, and indeed cycling it seems can continue to help facilitate it. Thus, as Cupples and Ridley go on to note, 'sustainability initiatives, rather than simply resisting or contesting neoliberal capitalism, can also work with and extend many of its individualising and totalising assumptions' (Cupples and Ridley 2008: 256).

A further point worth highlighting in relation to earlier cycling scholarship is that it presumes a homogenous cycling subject who will respond to pro-cycling measures in roughly the same way. Primarily, this can be seen in the statistics used to make the case for increases in cycling. For example, Lumsdon and Tolley (2001) note that, 'most trips in urban areas are very short. For example, nearly 75% of all journeys are under 8 km in length; nearly half are under 3.2 km and 30% under 0.6 km. Even in the case of journeys undertaken by car, 60% are below 8 km in length (Lumsdon and Tolley 2001: 294). Accordingly, Lumsdon and Tolley conclude that these distances are cyclable by the majority of the population. There is a sense here that cycling is a universal good and that everyone is equally capable of cycling and responding to incentives to cycle. A very similar message is delivered in Tolley (1996): 'some 40% of all car trips in Britain are under 1.6 km in length: most of these could switch to bicycles given suitable encouragement' (Tolley 1996: 215).

This homogenous user is also evident in the national archetypes drawn upon to demonstrate cycling best practice. Lumsdon (2000), for example, draws on best practice from Denmark and the Netherlands as well as further afield to demonstrate how such infrastructure can best encourage people out of their cars (Lumsdon 2000: 366). Similarly, Lumsdon and Tolley (2001) cite a number of paradigmatic places (Denmark, Germany, Netherlands) as cycling success stories, mainly because they have high overall numbers cycling. What is pertinent here with regard to difference is that there is no sense of who is or isn't cycling in these places; it is the overall numbers that matter as a measure of success and something to aspire to in the mainstreaming of cycling.

This lack of differentiation is also evident in the range of measures instituted to increase levels of cycling. This is quite striking in a paper by McClintock and Cleary (1996). Here they examine the effect of cycle facilities on the safety of cyclists in Nottingham (UK). There is recognition in the paper that cyclists are not a homogenous group, but the emphasis with regard to difference is placed upon age and ability as differentiating factors (74). When they discuss the safety of users and their views on cycle infrastructure, it is with regard to traffic safety and the ways in which poor cycle infrastructure can compound this, such as inadequate lane widths and poor visibility (69–70). There is no mention of social safety and gendered experiences of it in the set-up of the paper, nor is there discussion of this in the survey results reported. Indeed, the results highlight generic factors that affect safety – such as off-road routes making cyclists mix with pedestrians – rather than factors that might be specific to a particular group (72). This one-size-fits-all approach is also evident in Tolley (1996), where the mechanisms through which cycling levels can be increased are policy levers and exhortations to 'encourage modal shifting' (1996: 214). There is no sense that the ability of a person to respond to these mechanisms may be cut across by age, income, gender, or ethnicity (see, for example, Xie and Spinney 2018). What is evident is that the social is largely written out: it is assumed that cycling policies will be successful in recruiting users given suitable targets, infrastructure and awareness. Cupples and Ridley (2008) have referred to this as a potentially dangerous 'cycling fundamentalism' that overlooks issues of difference in the scramble to encourage cycling.

Whilst there is recognition of the needs for both 'hard' and 'soft' measures, there is an emphasis in research conducted in the 1990s and 2000s (and beyond) on planning and design solutions. For example, Lumsdon (2000) has been at pains to point out the ways in which cycling could contribute to lowering the harmful effects of transport in a tourism context. Like many other accounts of the time, there is an emphasis here on the extent to which new infrastructure, such as the National Cycle Network (NCN), could shift behaviour from car to bicycle. Designed primarily as a transport system, the 8,000 km UK National Cycle Network developed between 1995 and 2000 is described as 'a linked series of traffic-free paths, traffic-calmed and minor roads, connecting urban centres and the countryside throughout the United Kingdom' (Lumsdon 2000: 366). Likewise, Lumsdon and Tolley (2001) note considerable depressed demand for cycling, and suggest that releasing this will be a matter of comprehensive local strategy, the most important component of which will be building cycle tracks (Lumsdon and Tolley 2001: 295). McClintock's (1992) *The Bicycle and City Traffic* also places its emphasis here in seeking to assess patterns of bicycle usage and look at ways to increase it. Focused largely on the British and European context, as you would expect of a planning book, proposed solutions focus on fixes such as land use planning, traffic calming, home zones, etc., public transport integration, education and training. Within this there is an implicit assumption of a homogenous cycle user who will respond similarly to interventions. What is writ large in this literature is a mantra of, 'if we build it, they will come', with (somewhat understandably, given the lowly status of cycling at the time) no problematisation of who 'they' are: evidently the desire is to get more people cycling (primarily to work), rather than think about who might be excluded from such exhortations or why.

With cycling slowly growing in status and popularity, recent transport research has been more sensitive to difference, though still with a largely unquestioned focus on cycling as a force for good due to environmental and health benefits. Song *et al.* (2017), for example, cite the congestion, pollution, and physical inactivity resulting from automobility before outlining the potential of walking and cycling to address these challenges. Woodcock *et al.* (2007) discuss the detrimental effects of transport-related fossil-fuel emissions on the environment, with Spotswood *et al.* (2015) noting the negative consequences of automobility on human health, including obesity and physical inactivity. Cycling is presented in such accounts as a form of exercise enabling a reduction in a range of diseases, including cardiovascular, some cancers and type 2 diabetes (Yang *et al.* 2010). Similarly, Celis-Morales *et al.* (2017), Torjesen (2017) and Andersen (2017) all discuss the benefits of cycling to health of the population.

Within this body of work there is greater recognition that not everyone has the same ability to take up cycling due to a range of social factors. Kloof *et al.* (2014) conclude that a diverse range of factors, including ethnicity, gender, family circumstances, income and age, influence ability to cycle. Rietveld and Daniel (2004) note that cultural traditions and ethnicity play an important role in whether individuals are more or less likely to choose cycling as an alternative mode of travel. Indeed, as Junvaria (2018) has noted, Bangladeshi and Pakistani men in the UK tend to report lower levels of physical activity than other social ethnic groups, with Fischbacher *et al.* (2004) concluding that levels of physical activity were lower in all South Asian groups than the wider population. In relation to cycling specifically, Song *et al.* (2017) found that ethnic minorities and women were less likely to shift travel behaviour from car to active modes. Bowles Green Ltd (2008) specifically cite religion as an influencing factor on many Asian men and women, noting that husbands can feel ashamed or humiliated if wives and daughters are seen to participate in cycling or other physical activity (cited in Junvaria 2018).

This sits at odds to some extent with much of the literature, which tends to promote a standardised suite of measures to promote cycling, most notably, 'if we build it, they will come'. Panter *et al.* (2016), for example, found that new cycling infrastructure could increase cycling levels by as much as 80 minutes per week, with new and inexperienced cyclists most affected. As Pucher *et al.* (2010) conclude, a range of studies have concluded that dedicated bicycle infrastructure increases the quality and quantity of cycling. Similarly, studies by Dill and Carr (2003) and Parkin *et al.* (2008) found a correlation between higher levels of bicycle commuting and the presence of cycle infrastructure. However, as both Steinbach *et al.* (2011) and Pucher *et al.* (2010) conclude, despite a substantial body of evidence advocating cycle lanes as the solution, little is known about which are the most effective strategies at engaging different ethnic groups (and to some extent women) in cycling, given the additional cultural and social norms that govern behaviour.

It is evident from this discussion that the strategies used to conceive of place and practice within these approaches often forget that spatial patterns and processes are implicated within deeper economic and political structures that enable and constrain particular forms of action, an argument levelled and developed chiefly by Marxist geographers throughout the 1970s and 1980s. Certainly with few exceptions, older transport geography approaches have failed to take into account the different abilities of individuals – in the form of social identities such as class, race and gender – to accommodate or resist the structuring powers of the state (Massey 1991; Law 1999). Moreover, whilst city form has long been considered as a physical articulation of social relations and ideologies, with changes to their structure providing insight into broader political change (Hackworth 2007: 79), transport and its infrastructures in these accounts have been regarded as 'value neutral' rather than as an agent of capital accumulation and/or exclusion. A second criticism of rational approaches has been that they are determinist and reductive of people's experience and capacity as active agents, a critique developed by humanist geographers. As Bonham laments:

> Confining the study of urban travel to a story about transport has silenced a multiplicity of travel stories which spill out and are beyond the origin and destinations of each trip. Travel stories might be told in many different ways. They might emphasize the experiences of journeys, the social interactions (pleasant and unpleasant) which take place through (and outside) the journey, or the way in which racialised and gendered identities are worked upon and elaborated through the journey.
>
> (Bonham 2006: 71)

Cupples and Ridley (2008) confirm Bonham's point of view, noting that 'the obsession with cycle lanes can be seen as a will to produce a rational spatial order and leads to a neglect of affective and embodied dimensions of cycling' (Cupples and Ridley 2008: 259). More recent accounts, particularly those emerging out of the mobilities/transport intersection, have begun to take the embodied and less instrumental aspects of cycling more seriously. Before moving on to consider accounts of cycling that have originated from the mobilities turn, I wish to highlight a point regarding the performative nature of academic accounts reviewed so far. These literatures are concerned to show that cycling can be the solution to growing matters of concern such as local and global environmental problems, human health and economic efficiency. As such, not only do they not acknowledge cycling's potential role in the continuation of broader processes of capital accumulation, they are evidently performative of a particular mode of political economy and governmentality because they have been instrumental in trying to bring cycling back into urban economies in specific ways. I will elaborate upon this in the final section of this chapter.

Cycling in mobilities scholarship

In contrast to mobility-as-transport approaches, the mobilities turn has increasingly sought to understand movement as a set of practices with a variety of meanings, causes, constraints and scales. The mobilities turn, therefore, recognises that mobility and movement are entangled with relations of power and identity that previous approaches tended to either sideline or ignore altogether. The project of working these questions into new research has been taken up with vigour in both geographic inquiry and related disciplines of sociology, urban design, architecture, cultural studies and anthropology. Certainly, judging by the volume of work, mobility, it would now seem, is everywhere. The car has been approached from a variety of new angles, with work from Carrabine and Longhurst (2002), Delalex (2003), Dodge and Kitchin (2007), Edensor (2003), Featherstone (2004), Ferguson (2009), Gandy (2002), Merriman (2006), Miller (2001), Raje (2007), Sheller (2004), and Vaaranen and Wieloch (2001) all describing a multiplicity of in-car experiences from a variety of perspectives. Similarly cycling, which had until 2000 been largely confined to the fringes of academic enquiry, has seen a resurgence with work on the technology of cycling (Cox and Van de Walle 2012), bicycle activism (Aldred 2015; Furness 2005, 2007; Horton 2006, 2007), bicycle messengers (Ferguson 2017; Fincham 2006, 2007; Kidder 2005), digital economy and cycling (Bright Grayer and Brophy 2019; Briziarelli 2018; Shapiro 2018), urban cycling (Jones 2005; Spinney 2008, 2010; Stehlin 2014), public bike-share (DeMaio 2009; Strömberg 2017; Spinney and Lin 2019), road racing (Palmer 1996; Spinney 2006), identity (Aldred and Jungnickel 2014; Parkin *et al.* 2007; Skinner and Rosen 2007; Steinbach *et al.* 2011), and mediatisation (Hengst 2000).

Fincham (2004, 2007) and Kidder (2005) have both conducted ethnographic studies of bicycle messengers in the UK and the US respectively. Fincham focuses largely on the character of messenger work, its attendant risks and the ways in which these inform a messenger identity. Kidder's account highlights the production of a messenger 'style' through clothes, technology and the risks of the job. Both accounts point to the reworking of dominant meanings of the bike through specific embodied practices, suggesting that the meanings of cycling are more fluid and complex than previous accounts gave credit. More recently Lee (2018) has researched immigrant food delivery cyclists in New York to highlight the marginal status of this group and their invisibility in contemporary bike advocacy (125). Lee (2018) brings a more critical perspective to the study of cycling work and begins to link this to broader processes of migration and racialisation (125).

The work of Catherine Palmer (1996, 2001) and my own work (Spinney 2006, 2008, 2010) has drawn attention to the ways in which the practice of cycling is structured in relation to the bike, body and landscape to produce particular meanings of place. Palmer's (1996) ethnographic study of French racing cyclists focused particular attention on the technology of the bike, but did so in a way which framed the meanings of the technology at more discursive and mediatised levels rather than at the bodily scale. My own, much briefer, 'kinaesthetic' ethnographic work (Spinney 2006) with racing cyclists, whilst acknowledging the designed and discursive dimensions, has attempted to show how the meanings of technology are 'tested' and subsequently concretised through the practical conjoining of bike and body.

Often misrepresented as a form of instrumental mobility, cycling has been marginalised as a social and leisure activity in the urban landscape. The same reductionist logics that have framed the bicycle as a vehicle have served to exclude it from public and pedestrian environments because it legally 'belongs' on the roads and therefore must be too dangerous to safely mix with pedestrians. This preconception of what cycling should and should not be has resulted in a dearth of research relating to cycling as a form of urban leisure. In one of the few accounts focusing on this, Stevens (2007) suggests that people's definitions of play often move beyond the simplified rational and instrumental definitions of function materialised in the urban environment (1). As a consequence, Franck and Stevens (2007) point to the existence of what they term 'loose' spaces; spaces characterised by a lack of definition and which facilitate unplanned uses. The liminality of these spaces is used variously to transgress, to display, to party and to meet. As well as skateboarding and in-line skating, Stevens (2007) focuses in particular on the role that cycling has in (mis)using such spaces. Stevens notes how different urban cyclists use various edges and surfaces beyond what was intended by the designers (144). Forms of urban 'trick' cycling such as BMX and Trials riding in particular transgress spatial boundaries in the quest for heightened sensory experiences (particularly the kinaesthetic – the feeling of movement within the muscles and position of the body in space). Objects meant for static appreciation when used for hopping and jumping temporarily become something else. Stevens (2007) qualifies this, commenting that physical edges are used by riders because they afford 'sudden, thrilling exposure to shifts in height and speed' (144). Alongside Jones (2013, 2016), Stevens' account begins to link practices of cycling into broader political-economic processes, in this instance waterfront redevelopment. However, these accounts say less about exactly how such practices contribute to the life and vitality (and potentially profitability) of these areas.

Cycling, however, is not only meaningful in the context of the journey itself. As a politicised practice, many forms of mobility become meaningful to people because of the networks that surround them, in the form of subcultures and various forms of activism. Furness (2005, 2007) discusses some of the ways in which the bicycle has become meaningful to a disparate group of punks, anarchists, socialists and environmentalists, for whom ‘bicycles are not merely forms of transportation, rather, they are instruments of communication, sources of identity, vehicles for pleasure, and tools for technological, cultural, and political critique’ (Furness 2005: iv). As Furness goes on to suggest, the bicycle thus assumes a liminal status in the urban environment because it is more than simply transport; it is many things to many people. More recent accounts, such as those of Meador (2016), have begun to focus on diversity amongst bike activists. However, as Meador illustrates (though does not explicitly acknowledge), such groups still play a role in promoting biking (2016: 168) that does little to question how such alternative forms of mobility and their advocates come to be part of a governmentality of mobility (Cupples and Ridley 2008) through their performance and promotion of cycling.

Cycling and politics

As Baerenholdt (2013) has stated regarding mobilities scholarship, despite attention to everyday mobile social practices, ‘when dealing with issues of power, hegemony and social order, mobility studies are rather vague’ (Baerenholdt 2013: 21). In relation to cycling, for example, we’ve seen many papers (my own included) which have looked at what cycling as a practice means to those who do it as a sensory engagement (Jones 2005, 2012; Jungnickel and Aldred 2014; Nixon 2012; Spinney 2006, 2007); as a subcultural practice (Fincham 2006; Furness 2005); as a gendered practice (Cresswell and Uteng 2008); and as a response to health and environment concerns (Bohm *et al.* 2006; Bonham 2006; Button and Nijkamp 1997; Newman *et al.* 1995). The likes of Nixon (2012) and Fincham (2006) have both produced accounts that tie the embodied practice of cycling into the politics of urban mobility and energy use. Jones (2005) and Cupples and Ridley (2008) have both produced accounts that seek to tie effects arising from cycling into its broader promotion and spaces. Whilst insightful, these accounts support the idea that if mobilities scholarship has been viewed as apolitical, it is because its examinations of micro-politics has largely failed to relate these to wider political-economic processes.

Zuev *et al.* state that:

> with market societies dependent upon individually autonomous but disciplined and dependable circulation for day-to-day functioning

> (Foucault, 2010; Rajan, 2006), mobility and its government (both top-down and bottom-up) becomes a key locus of politics and power. Which forms of mobility are accepted, actively adopted and/or legitimated and which penalized, neglected and/or rejected is thus a key element of the trajectory of socio-technical change.
>
> (Zuev *et al.* 2018: 6)

In line with this observation, more recent cycling scholarship has begun to provide critical accounts that link cycling subjects with broader issues of urban governance (Golub *et al.* 2016; Spinney 2016; Spinney and Lin 2019; Stehlin 2014).

In one of very few papers linking increases in cycling with broader urban processes, Danyluk and Ley (2007) find an over-representation of cycle commuting in gentrifying areas of Canadian cities (see also Braun *et al.* 2019 in US cities). Likewise, Pucher and Buehler (2011: vii) confirm that the majority of cycling in US cities occurs in gentrifying areas. More recently, papers by Hoffman and Lugo (2014) and Stehlin (2014) have begun to make links between cycling and gentrification in a more overtly political way. Stehlin, for example, highlights the fact that neighbouring poorer districts (and those who cycle in them) have none of the cycle-specific facilities evident in gentrifying areas (2014: 30). Similarly, Hoffman and Lugo (2014) discuss the relationship between the creative class and bicycle infrastructure, arguing that despite good intentions, the emerging geography privileges access for some whilst also being implicated in urban redevelopment and competitiveness strategies.

One of the first books to try collect a range of broadly political-economic scholarship on cycling in a coherent volume was Golub, Hoffman and Lugo's (2016) *Bicycle Justice and Urban Transformation: Biking for all?* In this volume they argue that whilst the cycling renaissance in the US context can be seen as the preserve of the white middle class who ride from choice, the majority of cyclists are Latinos who ride out of necessity (2016: 2). As such they highlight both its emancipatory and oppressive potentials. The key goal of Golub *et al.* is to draw attention to those cyclists – women, the young, old, disabled, those on low income, and ethnic minorities – who have been overlooked in previous scholarship and connect this with larger questions around health, wealth and security (2016: 2).

Lam (2018) focuses specifically on the London borough of Hackney to question why the high cycling borough has so few women cycling (roughly one-third of cyclists) (Lam 2018: 116). Here – echoing Cupples and Ridley (2008) – she concludes that, amongst other things, the intent to mainstream cycling has rendered the Hackney Cycling Campaign partially blind to the need for diversity (120). This chimes with the findings of the earlier literature review section, which found that the desire to put cycling back on the agenda has led, as Stehlin notes, to a 'reductive understanding of what facilitates a vibrant urban cycling culture' (2014: 36). Lam's paper, like those of

Aldred and Jungnickel (2014), goes a long way to act as a corrective to earlier accounts that ignored issues of gender and difference more widely. The work of Golub *et al.* (2016) and Lam (2018) builds upon the work of Stehlin (2014) and sheds light on many of the reasons that the cycling resurgence manifests amongst certain places and populations, and seeks ways to increase diversity. However, it has less to say regarding how cycling has become situated within broader circuits of capital and 'produced' the renaissance, or indeed the extent to which cycling has been co-opted.

Of the literatures that really begin to dig into the relationships between cycling and broader political-economic and socio-technical processes is work on the gig economy. As services like Deliveroo and Uber Eats have expanded rapidly using low-wage mobile workers, so too has academic research. The likes of Briziarelli (2018) and Bright Grayer and Brophy (2019), for example, have examined the ways in which digitalisation has impacted on delivery riders in the context of post-Fordist capitalism. Labour–management interfaces used by companies such as Deliveroo, Uber Eats, Foodora or Glovo enable a labour process characterised by an extreme degree of employee surveillance and data extraction; utilisation of platform channels for management propaganda; algorithmically based managerial communication; the channelling, obstruction and suppression of worker-to-worker communication and even outright disconnection from the app (Bright Grayer and Brophy 2019; Vandaele *et al.* 2019). This body of work has begun to highlight not only the intersections between virtual and physical mobility and their impacts, but also the emerging strategies and tactics of resistance.

One of the few accounts that attempts to link UK cycling policy with a broader neoliberalisation is Aldred (2012). Aldred argues that whilst concepts such as 'hollow state' and the 'responsible individual' have gained traction within social policy, their use in transport policy is less evident (96). She goes on to argue that by positioning cycling in relation to health and environment debates (rather than just as 'transport'), the individual becomes 'urged' to act as part of an agenda of responsibilisation (2012: 99). Whilst such a reading begins to link cycling policy to neoliberal and bio-political modes of governmentality, it sees UK cycle policy as impacted by neoliberal doctrine rather than conceptualising it as a principle mode of neoliberal government. Cupples and Ridley (2008) go further to specifically refer to cycling policy and its attendant materialities as 'part of broader transformations that have taken place in advanced neoliberal societies that emphasise a notion of the active, self-managing and responsible citizen both on an individual and community level' (Cupples and Ridley 2008: 256).

In questioning the uneven geographies and moralities behind cycle promotion, this nascent literature at the intersection of mobility studies and transport geography is increasingly linking particular manifestations of mobility with broader political-economic processes of capitalist accumulation. Writing in 1989, Harvey asked whether the shift to urban entrepreneurialism would see 'even the most resolute and avantgarde municipal socialists … in the end, playing the capitalist game and performing as agents of discipline for the very processes they are trying to resist' (Harvey 1989: 5). It is quite possible that those concerned with mainstreaming cycling (I include myself in this group) may well just be those municipal socialists. In service to a mission to move cycling from an alternative and niche past-time to a mainstay of urban transport, activists and academics can end up 'playing the capitalist game', not because they are promoting cycling per se, but because their vision becomes co-opted into a myopic agenda that results in an exclusive and narrow vision of cycling focused on enhancing urban economic competitiveness. Whilst some more recent accounts have successfully broadened the vision of cycling, for many it is arguably still a vision that doesn't question the role of cycling in recuperating existing (unsustainable) growth patterns; or in helping to incorporate social cycling practices into spheres of capital accumulation. Certainly, as Stehlin has argued, contemporary cycling advocacy and practice must be situated critically within broader processes of capitalist urbanism (Stehlin 2014: 22). Whilst there is evidently a growing desire to understand cycling within broader political-economic frameworks, at present there is no coherent theorisation of this.

Theorising new urban geographies of cycling

A modest hope for this book is that it can begin to remedy this absence by providing an account that fuses insights from political economy, governmentality and mobility studies. Such an account is broadly post-structuralist in its outlook. In particular I am keen to provide an account that allows for feeling and sentient human bodies, rather than the passive, homogenous, disembodied user found in early political-economy and transport geography; one that pays attention to the ever-closer links between physical and virtual mobility; one which recognises the importance of materiality and the 'missing masses' (Latour 2008) in mobility projects; and one which also recognises cycle promotion as an often subtle and uncertain accomplishment achieved through multiple 'agencements' (Çalışkan and Callon 2009). Ultimately in bringing together a more grounded mobilities perspective and a political-economic one, such a theorisation seeks to 'connect questions of government, politics and administration to the space of bodies, lives, selves and persons' (Dean 2010: 20). In the following section I outline three sets of theory that underpin the ensuing chapters.

Urban political-economy and mobility

As Harvey (1991) has persuasively argued, we are witnessing a transformation in the political economy of twenty-first century capitalism, and it is incumbent upon us to understand what these changes are and how they take route (121). Writing in 1991 on the time-spaces of the contemporary city, Harvey argues that because of its relation to labour, money has no meaning independent of time. As a result, it is 'always possible to pursue profit (or other forms of advantage) by altering the ways time and space are used and defined' (1991: 229). Harvey notes that the 'battle over minutes and seconds' has been and continues to be fought across various territories: intensity of work schedules; the life course with regard to retirement rights; length of the working week/day; right to holidays, to name a few (231). Whilst Harvey does not explicitly mention it, I would add that mobility in general, and the journey to work in particular, are also increasingly problematised battlegrounds.

Where Harvey discusses mobility, it is with regard to the circulation of goods and labour. Any complex system of production requires spatial organisation and thus mobility (movement through time and space) to overcome it (1991: 229). Given that the time of production and time of circulation in exchange make up the 'turnover time of capital', 'efficiency of spatial organisation and movement is therefore an important issue for all capitalists' (Harvey 1991: 229). This is because the faster capital can be circulated, exchanged, and surplus value realised, the greater the profit will be (ibid.). There is for Harvey an imperative to accelerate turnover time beyond the 'social average' – e.g. the pace at which we might conduct our lives otherwise – thus promoting a trend toward faster average turnover times (ibid.).

Harvey theorises mobility as enhancing capital accumulation in two key ways: first, as far as physical goods are concerned, mobility is required to realise the surplus value in goods by bringing them to market – technological advances that speed up mobility serve to annihilate space through time. The benefit of this to the capitalist is that the time capital spends in circulation (as unproductive) is reduced. Second, these same technological advances in mobility can serve to bring labour together with capital more quickly and with less friction in order to enhance productivity: 'the more mobile the labourer, the more easily capital can adopt new labour processes and take advantage of superior locations. The free geographical mobility of labour power appears a necessary condition for the accumulation of capital' (Harvey 1982: 381). As a consequence Harvey points out that 'massive amounts of capital and labor have been invested in the sorts of immobile fixed capital we see in airports, commercial centers, office complexes, highways, suburbs, container terminals' (Harvey 2003a: 28). Indeed, the airport is just one example of a proliferation of innovations, including 'the railroad and the telegraph, the automobile, radio, telephone, the jet-aircraft … and the recent telecommunications revolution' dedicated to removing spatial barriers which have rendered the history of capitalism a very 'geographical affair' (Harvey 1991: 232). Accordingly, Harvey has long argued that mobility is central to productivity and acknowledges that many of what he has termed spatial fixes[1] are more accurately spatio-temporal fixes precisely because they are focused on mobility infrastructures to provide solutions to contradictions in capital accumulation (2003b: 87). Indeed, Harvey posits that any actor who can influence investments in mobility is likely to reap material rewards (Harvey 1991: 233).

There are, however, numerous barriers to this speeding up with whole industries (e.g. logistics, process management) dedicated to reducing such barriers from assembly line processes to planned obsolescence in production (Harvey 1991: 229). A key problem highlighted by Harvey is that the physical infrastructures upon which mobility relies are 'immobile in space and highly vulnerable to place-specific devaluation' (Harvey 1982: 380). As Harvey points out, 'capitalism increasingly relies upon fixed capital (including that embedded in a specific landscape of production) to revolutionise the value productivity of labour, only to find that its fixity (the specific geographical distribution) becomes the barrier to be overcome' (Harvey 1982: 394). In many cities across the world, the 'success' of particular forms of mobility (notably the car system) has resulted in the circulation of capital and labour slowing down to the point where these systems of mobility reduce the rate of capital accumulation because goods and workers are stuck in traffic. Moreover, it is evident that whilst sedentary forms of mobility such as the car, bus[2] and train have historically increased circulation times of labour, they have lowered its productivity in other ways by producing less healthy and sedentary labourers (cf. Bonham 2006; Horton 2006).

Entrepreneurial cities and images of place

Beyond this, the work of mobility can also be valorised in relation to place images. Harvey has long argued that states (at the national, regional and city scale) are – as economic entities – increasingly in competition with each other (Harvey 1991: 259). Harvey argues that as capitalists become increasingly sensitised to the different qualities of places, it becomes possible and desirable for the governors of those places to alter them in ways that make them more attractive to mobile capital and development (Harvey 1991: 295). Accordingly, the active production of special qualities of place becomes an important resource 'in spatial competition between localities, cities, regions, and nations'; moreover, 'corporatist forms of governance can flourish in such spaces, and themselves take on entrepreneurial roles in the production of favourable business climates and other special qualities' (Harvey 1991: 295). Thus we increasingly see the need for cities to appear creative (Mould 2018), sustainable (Bulkeley 2010), liveable (Southworth 2005), and healthy (Davison and Curl 2014) in order to reassure capital that it will see a return on investment.

Whilst this competition takes material and cultural form with the likes of creative parks, cultural quarters, river developments and urban markets, cities (but also nations) compete at the level of images (which have in themselves become commodities) because these allow instantaneous communication and enable the projection of their values (such as innovation, respectability, sustainability), authority and power across space (Harvey 1991: 288). In short, they enable places to establish an identity in a crowded market of places (ibid.). Importantly, 'given the ability to produce images more or less at will, it becomes feasible for accumulation to proceed at least in part on the basis of pure image production and marketing' (Harvey 1991: 289). As Book *et al.* (2010) note, 'Strategies to attract businesses, investors and people (inhabitants as well as visitors and tourists) to a place are a vital part of what is usually called place marketing. One central element of place marketing is image building' (383). Such images often co-opt 'oppressed groups' and their struggles to assert identity, and seek to convert this to commercial advantage (Harvey 1991: 289). However, as Harvey and others have been at pains to point out (cf. Stevens and Dovey 2004; Mould 2018), as policy becomes more mobile (McCann 2011; Wood 2014), such competition has led to a 'serial monotony' where particular tropes and themes become repeated as places glean 'best practice' from each other and copy ideas and schemes (Harvey 1991: 295).

Into this mix of special qualities of place we can add mobility: the ability to differentiate place by how the inhabitants move around is increasingly seen as a measure of how progressive and liveable a place is. In the absence of any real sustainability initiatives, and an unwillingness to tackle the continued growth of automobility, active mobilities like cycling can act as a symbol and image of sustainability, liveability, health and even creativity (cf. Florida 2000; Stehlin 2014) to suggest that environmental concerns are being addressed and that capital should locate there. As Harvey notes, whole cultures and sign systems can be co-opted as part of the specialised imaging of place (1991: 299). Indeed, it is 'precisely because capitalism is expansionary and imperialistic [that] cultural life in more and more areas gets brought within the grasp of the cash nexus and the logic of capital circulation' (Harvey 1991: 344).

Harvey notes that fixed material forms cannot be easily changed and the capitalist must either wait until their useful life is over before replacing them, or engage in creative destruction, which stands to prematurely lose a lot of the value tied up in the asset (Harvey 1991: 230). Consequently, many governments may prefer fixes (such as image building) that do not entail either large investments in new infrastructure or the premature loss of value in existing infrastructure: there is a tendency instead to rely upon shifting social and cultural norms and behaviours (Jones 2013) in service to image building – a strategy that largely operates within existing spatial structures with little or no destruction required. It also has the additional potential to 'activate' sedentary commuters/workers, thus potentially 'fixing' population-level health problems.

Fixing the mobile subject

The issue for municipal governments then is not just one of decreasing circulation time through fixing mobility, but also enhancing human capital and city image through mobility in other ways. Harvey implies as much when he states that, 'in so far as many aspects of physical and social infrastructure are fixed in space, the problem of geographical mobility is converted into one of transformation of the social and physical environment within which other forms of capital circulate' (Harvey 1982: 406). The role of particular fixes in urban governance then 'has thus become much more oriented to the provision of a 'good business climate' and to the construction of 'all sorts of lures to bring capital into town' (Harvey 1989: 11).

It follows from this that fixes instituted to enhance the productivity and competitiveness of the city will not only be characterised by their ability to directly reduce the circulation time of capital and labour, but also in their ability to enhance competitiveness in other ways. In line with this, Foucault (2010) has argued that economic growth cannot be explained through the speeding up of labour or capital alone. Rather, explanation must also turn to the ways in which human capital has been augmented and how this contributes to increases in productivity (232). As Cerny (2006: 690) goes on to state, the competitive state[3] is not only one whose economic policies focus on promoting competitiveness, but one which attempts to incorporate competitiveness into everyday social reproduction. Accordingly Foucault argues that modern policies of growth are 'focused precisely on one of the things that the west can modify so easily, and that is the form of investment in human capital' (2010: 232). As Guthman (2009) has argued, the material contradictions of capitalism are increasingly resolved in bodies (1114).

Whilst often criticised for downplaying 'extra-economic' factors such as human capital, Harvey is not blind to this. Writing in 1989, Harvey theorises a shift from managerial forms of urban government to entrepreneurial forms of governance as cities must increasingly prove their competitive worth on a global stage. Harvey suggests that entrepreneurialism essentially manifests in four key ways in municipal governance, but the most salient here is increasing competitiveness through the creation of new forms of labour power (Harvey 1989: 9). Harvey discusses the de-skilling and reskilling of the worker, arguing that when the skills of the labourer slow down turnover, workers must be reskilled to speed it up again. As a result, adaptable and flexible 'entrepreneurial' workers are crucial to such development (ibid.).

This logic is also applicable to the non-wage work of mobility, and particularly of the commute: if the commute is slowing down circulation times by holding up goods or workers or rendering them less productive (for example through poor health), the subject must be 'reskilled' to improve these. Hence in the modern city, the reskilling Harvey discusses is important not just in the context of waged labour, but in spheres of social reproduction. In the modern city this means a labour force that can gain the skills to move in ways that reproduce labour as a more productive form of human capital. Within this framing, mobility is problematised because increasingly our social reproductive activities outside of the sphere of paid labour (but contributing to its reproduction), such as the commute, are seen to have a profound impact on the accumulation strategies of towns and cities for the reasons already discussed. In extending the logic of entrepreneurialism firmly into mobility, and particularly social reproductive mobility, what we see according to Foucault is an extension of market rationality to domains that are not primarily economic (2010: 323) through a bio-politics that seeks to problematise everyday practice according to pervading political rationalities.

Accordingly, a key problem that Harvey identifies is the need to exert 'sufficient control over the way labour power is deployed to guarantee the addition of value in production, and hence, positive profits' (1991: 122). Whilst Harvey discusses this in the context of wage labour, it is again easily extended to the sphere of non-wage labour, for example, the work we expend on getting to and from work. Indeed, Harvey goes on to state that the issue is one of converting the 'capacity to do active work into a labour process whose fruits can be appropriated by capitalists' (1991: 123). Importantly in this understanding, the work of mobility is not necessarily commodified for direct profit; rather, there is a focus upon areas of life that can potentially detract from productivity (pollution, obesity) in order to recuperate the city and its workforce back to a more maximal productivity.

Mobility has long been a way of governing populations in that it is a sphere of life where the social is rendered economic – i.e. its ontological status is reimagined as one where certain elements are foregrounded and others backgrounded in order to maximise economic efficiency. For example, attempts to encourage particular forms of mobility such as cycling and walking are increasingly couched with reference to improving efficiency and health, both of which require the foregrounding of these qualities of cycling and the backgrounding of many other qualities. In such readings, what becomes important about mobility is not its sensory character, its potential to bring people together, to create social meaning, or any enjoyment we might get from it; but rather its ability to produce citizens who are more efficient and less of a burden on the state.[4] Whilst such inducements are often aimed at the commuter – already a form of mobility central to economic reproduction – they may also be aimed at other social reproductive journeys such as shopping, leisure and journeys to school.

A key contention here is that mobility in the realm of social reproduction (e.g. commuting and leisure) can be a source of surplus value, what Zuboff (2019) – in the context of data capitalism – calls 'behavioural surplus'. Whilst Marx imagined that capitalism fed off labour to create a surplus, Zuboff argues that modern variants of capitalism feed on every aspect of human experience (2019: 9) as a source of surplus. When a person decides for whatever reason how to get to work, they engage in a behaviour that ostensibly satisfies their 'own' requirements: they get to work quicker, more comfortably, they save money, they have more fun doing it, etc. However, this behaviour can in some circumstances generate a surplus that is of value to a third party. To clarify this, it is apposite to refer back to

Marx when he argues that surplus value is the additional value generated in excess of labour and production costs (Marx 1990: 307). Of course in the example of commuting, the 'worker' is not at work and is not being paid. Yet depending on how a matter of concern is framed, cycling (and indeed other forms of movement) can be generative of surplus because of the multiple qualities of the practice, and therefore multiple ways in which the behaviour can be valued. Again, Marx sheds some light on what is happening here, by distinguishing between labour-time worked and labour power. Labour power is only a potential, and it is only through putting the labourer to work that this potential is transformed into actual labour. However, the wage of the labourer does not reflect the full labour power of the individual because the individual has the capacity to undertake many forms of work, only some of which are useful – and therefore remunerated – by the capitalist (Marx 1990: 283–294).

Following this logic, it is the additional capacity of the mobile subject to do work (let's say when cycling to work or the shops) that is valued because the value produced is beyond what the cyclist requires to meet their own (or the capitalist's) goals (e.g. in getting from A to B). Hence certain forms of mobility in certain contexts have greater labour power for a given labour-time worked. As this suggests, surplus value can become a proxy for the social productivity of the working population because it represents value produced by labour beyond that required to reproduce itself (Marx 1990: 293).

So in this sense, the social domain can be seen as an important arena for value creation beyond the realm of waged labour because of the capacity to engage in economically valuable work even when one is not 'labouring'. In line with Zuboff's observations on surveillance capitalism, it would appear that whilst 'industrial capitalism thrived at the expense of nature', new modes of accumulation thrive 'at the expense of human nature' (Zuboff 2019: 11–12). Importantly, the labour power of the mobile subject is not the object of a monetary value exchange in that they 'do not establish constructive producer-consumer reciprocities' (Zuboff 2019: 10). Rather, I suggest under these new terms of engagement, certain kinds of mobility are promoted and accommodated to entice users into 'extractive operations in which our personal experiences are scraped and packaged as the means to others' ends' (ibid.).

Mobility, technology and behavioural surplus

Writing in 2006, Sheller and Urry laid claim to a broad 'new mobilities paradigm' and set out a diverse agenda for mobilities researchers. One of the themes they highlighted was the need to study the intersection of virtual and physical mobility. They argue that mobilities research should question how such intersections are 'mobilised, or performed, through ongoing socio-technical practices, of intermittently mobile material worlds'

(Sheller and Urry 2006: 211). Indeed, they argue that hybrid mobility (between physical and virtual) should be a key area of study and a number of mobilities scholars have contributed significantly to our knowledge of these intersections.

Writing in 2009, Bratton argued that with the advent of the smart phone, our mobility is shifting from mechanical to informational, suggesting that the car is in decline and that the smart phone is the object in ascendancy (92). As Bratton elaborated, the smart phone is not replacing but augmenting and facilitating physical mobility. However, the majority of what the smart phone does is hidden to the user due to the nature of the pervasive computing revolution (Dodge *et al.* 2009; Galloway 2004) – where computers and software become embedded into our everyday lives in such a way that they are rendered invisible. This embeddedness leads Dodge *et al.* (2009) to ask a number of questions, perhaps the most salient of which is to ask the extent to which pervasive forms of computing challenge agency and democracy and produce/exacerbate inequalities (2009: 1284) because we have so little idea of what these devices are doing 'behind the scenes'. Indeed, as Williamson argues, our movement is increasingly animated and directed by black-boxed algorithmic processes (2015: 135).

Millington describes how a variety of software apps (such as fitness tracking, location-based social media, and mobile gaming apps) increasingly harness the possibilities available from the intersection of physical and virtual mobilities for individuals to quantify and know themselves (2014: 480). However, alongside the ability of the user to 'know' themselves comes the ability of corporations to know, commodify and share everyday user practice by gathering and selling data on their habits and preferences (Williamson 2015). To date Zuboff (2019) has published the most convincing and complete account of this phenomena in *Surveillance Capitalism*. Her focus here is largely on the big search and social media giants like Google and Facebook. She makes a convincing argument that uses and updates Marxist political-economy to argue that surveillance capitalism utilises the collateral traces of online behaviour to constitute a behavioural surplus (Zuboff 2019: 78). This is then combined with other data and monetised in the form of 'predictive products'.

However, Zuboff argues that whilst scale and scope of data are important for the manufacture of high-quality predictive products, 'the surest way to predict behavior is to intervene at its source and shape it. The processes invented to achieve this goal are what I call *economies of action*' (Zuboff 2019: 193–194). Zuboff goes on to argue that this requires the implementation of machine-based 'architectures of extraction' in the real world that 'weave themselves into the fabric of everyday life until they are indistinguishable from it, but with a twist. Now they operate in the interests of surveillance capitalists' (Zuboff 2019: 193–194). Accordingly any political economy of mobility must now take into account the new ways in which virtual and physical mobility are coming together to offer users new ways of moving around, new ways of knowing themselves and ultimately new opportunities for interested parties to extract behavioural surplus. Congruent with this, mobility as a service (MaaS) and in particular MaaS operated by private companies (such as dockless public bike-sharing) becomes a concern because of its potential to extract and render the behavioural surplus of mobility in a monetisable form.

Whether or not our personal experiences of mobility yield surplus value – are valorised – or not remains uncertain and is dependent upon what are framed as particular social problems and solutions at a given time. Certainly as Marx elaborated, valorisation is dependent on labour power being expended in a useful (productive) manner, but the realisation of surplus value ultimately depends on a separate issue process of monetrary exchange (Marx 1990: 300). Harvey states that 'value and meaning are not inherent in any spatial order … but must be invoked' (1991: 216). Clearly, as Zuboff's work highlights, the realisation of surplus value depends on the presence of technologies able to extract and process that surplus value. More importantly, however, it also depends upon a political-economic context that renders surplus meaningful and therefore is deemed to have value (e.g. be able to contribute to mitigating a particular matter of concern such as circulation times, worker productivity, city image or environmental degradation). Certainly we see that at different points in time, certain forms of mobility are problematised and devalorised as the cause of matters of concern whilst others are valorised as part of the solution (see, for example, Cupples and Ridley 2008). Hence the realisation of surplus value in the realm of socially reproductive mobility is dependent upon it addressing key governmental concerns such as economic efficiency, global and local pollution, city image and human health.

Labour control is essential to the process of valorisation in the sphere of social reproduction because it provides a framework through which any surplus can be interpreted as valuable and that can galvanise and provide a framework of meaning through which citizens can reskill themselves. Harvey argues that in order to produce labour (or in this instance citizens more generally) with the requisite skills to be mobile in appropriate ways requires the disciplining of labour power in the form of labour control: 'a mix of repression, habituation, co-option and co-operation, all of which have to be organised … throughout society at large' (1991: 123). This entails the social control of physical and mental powers including the mobilisation of social sentiments (work ethics, national pride), and psychological propensities (e.g. search for identity through work) (ibid.). As such, considerable power accrues to those who can influence representations of mobility in ways that enable labour control and produce particular subjectivities. Foucault asks, 'How does the human body become docile and useful in the context of a particular social system?' (1995: 163–169). Here, Foucault is interested in 'the way in which people are invited or incited to recognize their moral obligations' (1984: 264, cited in Kristensen 2013: 18). Following this logic, Galloway suggests that in modern governmentality:

> control is exerted by inducing action rather than restricting it, or, more precisely, by 'curating' a networked terrain within which action is nurtured. In the society of control mobility is fostered inside certain strictures of motion, where openings appear rather than disappear, where subjects (or for that matter objects) are liberated as long as they adhere to a variety of prescribed comportments.
>
> (Galloway, cited in Krivy 2018: 19)

In effect, this is what Foucault tried to show in his inquiries of the mid-1970s: how certain forms of subjectivity are nurtured into existence instead of others; in relation to which rationalities are certain subjectivities represented as more legitimate, normal and desirable while others are marginalised or excluded? My argument here is that for capitalism, if a particular subjectivity is required to address a particular matter of concern, the subject and its associated practice (or more importantly a particular version of that practice) may become co-opted: what might start as a work-around or critique by the few (such as a resurgence in urban cycling) becomes adopted, adapted and mainstreamed. As Tyfield has argued, 'countercultural critiques of the 1960s and 1970s came to inform what emerged subsequently as the dominant management discourses of the 1980s and 1990s, in a classic example of an anti-capitalist-critique-turned-capitalist-manifesto' (2015: 532). Hence we must consider the fact that the subjects and practice of cycling are transformed through their insertion into circuits of capital.

Mobility and the new economic sociology: qualities, attachment, ordering and translation

What is particularly interesting in relation to cycling as a practice is that until very recently it has been considered 'extra-economic', or indeed 'anti-economic' in that it has been viewed as counterproductive and 'inefficient' with regard to processes of accumulation, and (often as a result) been appropriated as a symbol of opposition to contemporary capital accumulation symbolised by the car (Furness 2005). A key argument of this book as set out in the introduction is that whilst still marginal in most places, cycling is playing a more central role with regard to accumulation in relation to multiple matters of concern. The question thus arises of how a social practice like cycling is rendered economic. The discussion thus far in this chapter has started to outline this in relation to the valorisation of labour power to produce surplus value. Here I want to expand more upon 'how' the social practice of cycling is rendered economic using work from the new economic sociology, a literature concerned to understand processes that have generally not been considered in orthodox accounts of market exchange.

As McFall (2009) has argued, the insights from the new economic sociology are useful because they are concerned 'with how things, behaviours and processes are "rendered" economic' (Muniesa *et al.* 2007, in McFall 2009: 273). Intentionally engineered processes that bring the social into economic realms are important because, as Thoburn (2007) attests, 'it is clear that neither culture nor politics can be understood without an intimate attention to the way capitalist dynamics and imperatives infuse the social' (79). Thoburn goes on to suggest that we need to approach power and culture through an understanding of production 'considered as the patterning – or mobilization, arrangement and distribution – of rich social, technical, economic and affective relations' (2007: 80). Power in such a reading constructs subject positions which serve to bound choices. However, power is not prefigured as theories of 'Kapitalism' would suggest, rather it is achieved through the alignment of different actors:

> It is not so much that capitalism (or, 'the economy') determines all social relations, but that the conditions of social production are set in a fashion that encourages certain formations, and that are socially posed as, or arise from, economic problems.
>
> (Deleuze 1994: 186, in Thoburn 2007: 82)

Thus goals like economic efficiency in mobility systems are addressed not through 'mass formations' but rather through arrangements and networks of 'matter and function' which are configured (Thoburn 2007: 83).

Such thinking relates closely to my contention that capitalist dynamics are increasingly infusing cycling as a social practice. Through this lens we can see the production of particular variants of cycling culture as an attempt to create a market for cycling by attempting to attach users to particular variants of cycling (by making it relevant in particular ways) whilst detaching them from their existing mobility practices (including less 'useful' versions of cycling). There are two elements of this that I want to discuss in further detail.

The first of these is agencement: broadly speaking this describes the interaction of elements to form a new equilibrium, dependent on the goals of the stakeholders. Agencement is useful insomuch as it highlights the fact that many of the processes through which social practices are rendered economic are subtle, diffuse and gradual rather than explicit and clearly visible. Processes of economisation, whereby 'activities, behaviours and spheres or fields are established as being economic' (Çalışkan and Callon 2009: 370) rely on agencements of discourses, material environments, policies, technologies, user competencies and dispositions, etc., because it is through these that practices are aligned with goals and 'mobility markets' are created through which subjects can be assigned and learn roles.

Examination of agencements rests upon analysing 'specific processes of association and ordering, which connect what was previously detached'

(Van Der Duim 2005: 962). In particular there is a focus on the ways in which discourses, materialities and practices are more or less successfully fused together (Van Der Duim 2005: 962). As Murdoch has outlined, the goal is to understand how associations and relations emerge, and in so doing, how functions and roles of different objects, actors and places emerge alongside (Murdoch 1997: 331). For Van der Duim, analysis should be directed toward particular elements of a given network. First, people and bodies: mobility is performed by those being mobile and those organising that mobility; second, a variety of objects, media and technologies serve to structure, direct and configure mobility; third, mobilities are performed in spaces constituted through multiple relations that serve to shape possibilities and meanings (Van Der Duim 2005: 967–968). To study mobility in this way is to employ a semiotic definition of actors where form and attributes are acquired as a result of the ways in which they are related to each other (Van Der Duim 2005: 963). Related to this, agencements must be seen not as fully organised and strategic in the sense of a collective (an assembly of people who have decided to join some form of common organisation) but rather a 'collectif' characterised as an 'emergent effect created by the interaction of the heterogeneous parts that make it up' (Callon and Law 1995: 485).

Agencement is therefore an important concept because it describes the production of 'markets' or framings where the appropriate qualities of mobility can be both promoted and inculcated. Examples of elements that constitute a market in this regard can include cycling events, cycle lanes, design guidance, policy, legal frameworks, cycle training, etc. It is through these that users come to recognise and perform what the important elements of a practice are, and a framework is provided through which user practice is aligned with strategic goals in order that any surplus value in a practice can be put to productive use and therefore valorised. In short, it is through agencements that labour power is enhanced because this creates the conditions for subjects to 'reskill' themselves and for any previously extra-economic (surplus) value of mobility to be performed and extracted. Hence it is the extraction of value from processes occurring outside of waged labour that is relevant here and how it is made possible through such alignments.

Related to this, the second point I want to discuss in relation to mobility markets and economisation is that of qualities and their relation to particular framings and matters of concern. Drawing upon work in actor network theory (ANT), I argue that economisation is first and foremost a process of establishing and ordering qualities (Latour 2008; Law 2001; Hummel and Van der Duim 2016). According to Callon and Muniesa (2005), any processes of market-making are processes of qualification and requalification that continually align product features to consumer interests in order that attachment can take place (Callon *et al.* 2002: 200–201). The list of what Callon *et al.* term qualities is not fixed and thus, for any firm, 'the ability to modify the list of qualities is a strategic resource since it is a matter of positioning the good in the space of goods' (Callon *et al.* 2002: 200). These processes of qualification consist of three inseparable elements: objectification, singularisation and attachment-detachment, all of which are informed to a degree by processes of calculation (e.g. evaluate the most beneficial course of action). Callon *et al.* (2002) note the need for producers to define or 'singularise' the characteristics of a good in such a way that the product stands out. Consequently all agents involved in a market are engaged in a constant process of requalification, which Callon *et al.* suggest falls into two categories – organised (such as market research where economic agents are explicitly defined as being involved in the process), and the gradual and unconscious (such as the placement of goods on a shelf or a conversation with a family member) (Callon *et al.* 2002: 201). Callon *et al.* (2002) state that central to this process are 'qualification professionals' who 'constantly try to destabilize consumers, to extract them from routines and prompt them to re-evaluate the qualities of products, hoping that that re-qualification might be favourable to them' (Callon *et al.* 2002: 206). Thus, whilst Callon *et al.* theorise the consumer as thinking and active, they also suggest that in many instances the tools of calculation that are used by consumers to evaluate products are proposed by supply-side actors (Callon 2005: 6).

Processes of defining important qualities (including those of mobility) and ordering them are important because it is through attaching particular characteristics to practices and attempting to make them 'attractive' (van Der Duim 2005: 966) that often disparate entities and ideas are brought together and made comprehensible. Hence the ordering of qualities and resulting agencements refer to 'strategic and coherent notions about how mobility should be performed' (Van Der Duim 2005: 969). As Law goes on, 'modes of ordering are mini-discourses ... [defining] ... conditions of possibility, making some ways of ordering webs of relations easier and others difficult or impossible' (Law 2007: 10). As a result, modes of ordering are on the one hand discursive narrations in that they are representations and idealisations 'consisting of values, facts, or the desires of planners, entrepreneurs, tourists, and locals' setting out what mobility should look like (Van Der Duim 2005: 969); but they are also practised, in that mobilities themselves provide congruity regarding how mobility should be performed (Van Der Duim 2005: 969). In both cases, (attempted) integration is implied as discursive and practical modes of ordering are brought together in a given network. Thus the work of imbuing and foregrounding specific qualities of practice, and the creation of a mobility market, 'is a distributed effort involving both market professionals and the final user' (Araujo 2007: 214).

Whilst the work of Callon and collaborators has generally applied to retailers, marketers, manufacturers and consumers, we can just as easily see various state and non-state actors as 'qualification professionals' precisely because their goal is to detach citizens from their existing mobility routines and attach them to alternate forms of mobility (such as cycling) through a re-evaluation of qualities. Hence the ability to control qualities is key to any chance of success in rendering the social economic. Again with regard to mobility it is clear that certain qualities – such as low emissions, small spatial footprint, positive health impacts, positive green symbolism, etc. – have become more important because they have value in mitigating particular matters of concern such as climate change, obesity and city image. It is therefore important to understand that what we currently deem are the important qualities of cycling are not the only, or in any way 'natural', qualities of cycling; rather, they are the ones promoted as important by particular actors because of contemporary matters of concern. Given a different set of priorities, it may well be that other qualities such as sense of freedom or playfulness might be foregrounded. Hence whilst consumers are theorised as active and calculating individuals with regard to which qualities of a practice are important, many of their calculations are evidently framed by market professionals (academics, planners, policy-makers, activist organisations). Such a reading suggests that whilst the passive 'controlled' user depicted in consumption studies such as those of the Frankfurt school is no longer prevalent (Lupton and Noble 2002: 6), neither are consumer judgements about product quality free from influence.

Conclusions

By way of concluding this chapter I want to reiterate the key absences and insights both empirically and theoretically in order to contextualise the chapters that follow. First, it is clear that rich insights from transport geography in relation to cycling have tended to neglect issues of distribution and difference. This literature – in building an evidence base for, and foregrounding particular qualities of, cycling has actively promoted particular variants of cycling as more or less desirable, with a notable focus on the commute and the role cycling can play within this as a mode choice. Partly as a consequence, research in this vein has tended to prioritise the adult, white male commuter cyclist with much less attention paid to issues of gender, age, ethnicity and how these shape the behaviours and perceived qualities of cycling. Academic accounts keen to put cycling on the policy map have played a key part in defining the important qualities of mobility and therefore actively constructed narratives around cycling and what constitute appropriate performances of that mobility in relation to matters of concern (such as climate change, urban economies and public health agendas).

Second, I have highlighted a lack of research in relation to mobility and political-economy, particularly in relation to cycling. In many accounts, the resurgence of cycling in urban environments over the last 20 years has been uncritically engaged with as neutral or positive, rather than resulting from and reproducing particular economic and political orientations. Questions of why we see cycling manifest in urban economies, why certain variants of cycling and not others are evident, and what the implications of this might be for urban mobility and cities more broadly have rarely been asked. Only in some recent accounts such as those exploring the gendered and classed nature of cycling promotion have we started to see engagement with the idea that whilst cycling may be a boon to urban areas, its emergence has a particular geography that reflects and reproduces broader logics and inequalities of capitalist accumulation.

Finally, this chapter has sought to bring together work in mobilities, political economy, governmentality and economic sociology to provide a framework through which we might understand why and how we see cycling manifest in particular ways in particular places. In the first part I made an argument for seeing contemporary manifestations of cycling as a mobility fix because it seeks to address problems of accumulation not through the production of space, but primarily through the production of mobile subjects. In line with this, rather than see economy as over-determining and totalising, I have highlighted the links between particular matters of concern; the qualities of cycling that are foregrounded as a consequence; the agencies that bring them into being (and in turn bring new agencies into being); and the often subtle, contingent and uncertain agencements that operationalise and enable the performance of these qualities and agencies. In doing so, I have attempted to provide a framework through which we can understand the economisation of cycling as premised upon a valorisation of cycling that is dependent upon particular qualities of it being foregrounded. Importantly variations in these qualities are not inevitable in the forms they take but represent a response to broader political and economic matters of concern. In line with Cupples and Ridley (2008), my goal here is not to undermine cycling practices as a (sustainable) form of transport, but rather to analyse 'the construction of their logics' and to unsettle 'their taken-for-granted status in discourses and practices' (Cupples and Ridley 2008: 255).

Notes

1 In a number of books and papers, Harvey (1982, 1989, 1991) has argued that falling productivity leads firms/cities/nations to seek what he calls spatial and temporal fixes. Put simply, a spatial fix is the movement of capital to new places to take advantage of 'idle' labour supply in order to increase productivity and maintain growth. Complementing the spatial fix, Harvey also uses to a lesser extent the notion of 'temporal fix': instances where the speed of the circulation of capital is increased in order to maximise productivity.

2 Whilst a discourse dominates which positions these as sedentary, we might want to question the extent to which this is true. Webb (2014), for example, has demonstrated that those who use public transport are 25% less likely to be obese than those who don't (4).

3 It should be noted that the state in such a reading is increasingly a collection of municipal actors.

4 As Jessop (2002) states, however, it is important to note that the economic can never be over-determining in such formations. Given that the economic and extra-economic are always in relation, one cannot be solely determining even though it may gain dominance.

References

Adams, J. (2001) *The Social Consequences of Hypermobility*. RSA Lecture, 21 November. Available at: http://john-adams.co.uk/wp-content/uploads/2006/hypermobilityfor RSA.pdf [accessed 21 September 2013].

Aldred, R. (2012) Governing transport from welfare state to hollow state: The case of cycling in the UK, *Transport Policy*, 23, pp. 95–102.

Aldred, R. (2015) A matter of utility? Rationalising cycling, cycling rationalities, *Mobilities*, 5(1), pp. 686–705.

Aldred, R. and Jungnickel, K. (2014) Why culture matters for transport policy: The case of cycling in the UK, *Journal of Transport Geography*, 34, pp. 78–87.

Andersen, L.B. (2017) Active commuting is beneficial for health, *BMJ*, 357. doi:10.1136/bmj.j1740

Araujo, L. (2007) Markets, market-making and marketing, *Marketing Theory*, 7(3), pp. 211–226.

Aultman-Hall, L., Hall, F.L. and Baetz, B.B. (1997) Analysis of bicycle commuter routes using geographic information systems: Implications for bicycle planning, *Transportation Research Record*, 1578, pp. 102–110.

Axhausen, K., *et al.* (2000) Observing the rhythms of daily life: A six-week travel diary, *Transportation*, 29, pp. 95–124.

Baerenholdt, J.O. (2013) Governmobility: The powers of mobility, *Mobilities*, 8(1), pp. 20–34.

Barr, S. (2003) Strategies for sustainability: Citizens and responsible environmental behaviour, *Area*, 35(3), pp. 227–240.

Bickerstaff, K. and Walker, G. (1999) Clearing the smog: Public responses to air quality information, *Local Environment*, 4, pp. 279–294.

Bickerstaff, K. and Walker, G. (2003) The place(s) of matter: Matter out of place – public understandings of air pollution, *Progress in Human Geography*, 27(1), pp. 45–67.

Bohm, S., Jones, C., Land, C. and Paterson, M. (2006) Introduction: Impossibilities of automobility, *Sociological Review*, 54(s1), pp. 2–16.

Bonham, J. (2006) Transport: Disciplining the body that travels, *Sociological Review*, 54(s1), pp. 57–74.

Book, K., Eskilsson, L. and Khan, J. (2010) Governing the balance between sustainability and competitiveness in urban planning: The case of the Orestad Model, *Environmental Policy and Governance*, 20, pp. 382–396.

Bowles Green Ltd (2008) on behalf of Blackburn with Darwen Borough Council and Lancashire County Council, *Engaging Ethnic Minority Communities in Cycling*.

Bratton, B.H. (2009) iPhone city, *Architectural Design*, 79(4), pp. 92–97.

Braun, L., Rodriguez, D. and Gordon-Larsen, P. (2019) Social (in)equity in access to cycling infrastructure: Cross-sectional associations between bike lanes and area-level sociodemographic characteristics in 22 large U.S. cities, *Journal of Transport Geography*, 80, pp. 1–10.

Bright Grayer, S. and Brophy, E. (2019) *Platform Organizing*. Available at: https://notesfrombelow.org/article/platform-organizing [accessed 18 September 2019].

Briziarelli, M. (2018) Spatial politics in the digital realm: The logistics/precarity dialectics and Deliveroo's tertiary space struggles, *Cultural Studies*, 0(0), pp. 1–18.

Bulkeley, H. (2010) Cities and the governing of climate change, *Annual Review of Environment and Resources*, 35, pp. 229–253.

Button, K. and Nijkamp, P. (1997) Social change and sustainable transport, *Journal of Transport Geography*, 5(3), pp. 215–218.

Çalışkan, K. and Callon, M. (2009) Economization, part 1: Shifting attention from the economy towards processes of economization, *Economy and Society*, 38(3), pp. 369–398.

Callon, M. (2005) Why virtualism paves the way to political impotence: A reply to Daniel Miller's critique of the Law of the Markets, *Economic Sociology European Newsletter*, 6(2), pp. 3–20.

Callon, M. and Law, J. (1995) Agency and the hybrid 'Collectif', *The South Atlantic Quarterly*, 94(2), pp. 481–507.

Callon, M. and Muniesa, F. (2005) Peripheral vision: economic markets as calculative collective devices, *Organisation Studies*, 26(8), pp. 1229–1250.

Callon, M., Méadel, C. and Rabeharisoa, V. (2002) The economy of qualities, *Economy and Society*, 31(2), pp. 194–217.

Carrabine, E. and Longhurst, B. (2002) Consuming the car: Anticipation, use and meaning in contemporary youth culture, *Sociological Review*, 50(2), pp. 181–196.

Celis-Morales, C.A., Lyall, D.M., Welsh, P., Anderson, J., Steell, L., Guo, Y., … Gill, J.M.R. (2017) Association between active commuting and incident cardiovascular disease, cancer, and mortality: Prospective cohort study, *BMJ*, 357. doi:10.1136/bmj.j1456

Cerny, P.G. (2006) Restructuring the state in a globalizing world: Capital accumulation, tangled hierarchies and the search for a new spatio-temporal fix, *Review of International Political Economy*, 13(4), pp. 679–695.

Cervero, R. and Radisch, C. (1996) Travel choices in pedestrian versus automobile oriented neighbourhoods, *Transport Policy*, 3(3), pp. 127–141.

Cohen, M. (2006) A social problems framework for the critical appraisal of automobility and sustainable systems innovation, *Mobilities*, 1(1), pp. 23–38.

Cox, P. and Van De Walle, F. (2012) 'Bicycles don't evolve: Velomobiles and the modelling of transport technologies', in Horton, D., Rosen, P. and Cox, P. (eds) *Cycling and Society* (Farnham: Ashgate), pp. 113–132.

Crane, R. and Crepeau, R. (1998) Does neighbourhood design influence travel? A behavioural analysis of travel diary and GIS data, *Transport Research*, 3(4), pp. 225–238.

Cresswell, T. and Uteng, T. (2008) *Gendered Mobilities* (Farnham: Ashgate).

Cupples, J. and Ridley, E. (2008) Towards a heterogeneous environmental responsibility: Sustainability and cycling fundamentalism, *AREA*, 40(2), pp. 254–264.

Danyluk, M. and Ley, D. (2007) Modalities of the new middle class: Ideology and behaviour in the journey to work from gentrified neighbourhoods in Canada, *Urban Studies*, 44(11), pp. 2195–2210.

Davison, L. and Curl, A. (2014) A transport and health geography perspective on walking and cycling, *Journal of Transport and Health*, 1(4), pp. 341–345.

Dean, M. (2010) *Governmentality: Power and rule in modern society*, 2nd ed. (London: Sage).

Delalex, G. (2003) *Living on the Motorway*, Paper presented at the Alternative Mobility Futures Conference, Lancaster University. Available at: www.spaceandculture.org/2004/01/alternative-mobility-futures.php [accessed January 2005].

DeMaio, P. (2009) Bike-sharing: History, impacts, models of provision, and future, *Journal of Public Transportation*, 12(4), pp. 41–56.
Dery, M. (2006) 'Always crashing in the same car': A head-on collision with the technosphere, *Sociological Review*, 54(s1), pp. 223–239.
Dill, J. and Carr, T. (2003) Bicycle commuting and facilities in major US cities: If you build them, commuters will use them, *Transportation Research Record*, 1828, pp. 116–123.
Dodge, M. and Kitchin, R. (2007) The automatic management of drivers and driving spaces, *Geoforum*, 38, pp. 264–275.
Dodge, M., Kitchin, R. and Zook, M. (2009) Guest editorial: How does software make space? Exploring some geographical dimensions of pervasive computing and software studies, *Environment and Planning* A, 41, pp. 1283–1293.
Edensor, T. (2003) M6 junction 19–16: Defamiliarising the mundane roadscape, *Space and Culture*, 6(2), pp. 151–168.
Featherstone, M. (2004) Automobilities: An introduction, *Theory, Culture and Society*, 21(4–5), pp. 1–24.
Ferguson, H. (2009) Driven to care: The car, automobility and social work, *Mobilities*, 4(2), pp. 275–293.
Ferguson, J. (2017) Discreet to excrete in the concrete jungle: Women bike messengers and their inventive urban strategies in three US cities, *Gender, Place and Culture*, 24(1), pp. 85–96.
Fincham, B. (2004) *Bicycle Couriers: Identity, risk and work*, unpublished PhD thesis (University of Cardiff).
Fincham, B. (2006) Bicycle messengers and the road to freedom, *Sociological Review*, 54(s1), pp. 208–222.
Fincham, B. (2007) 'Bicycle messengers: Image, identity and community', in Horton, D., Rosen, P. and Cox, P. (eds) *Cycling and Society* (Aldershot, UK: Ashgate), pp. 179–196.
Fischbacher, C.M., Hunt, S. and Alexander, L. (2004) How physically active are South Asians in the United Kingdom? A literature review, *Journal of Public Health*, 26(3), pp. 250–258.
Florida, R. (2000) *The Rise of the Creative Class and How It's Transforming Work, Life, Community and Everyday Life* (London: Basic Books).
Foucault, M. (1995) *Discipline and Punish: The birth of the prison*, trans. Alan Sheridan (New York: Random House).
Foucault, M. (2010) *The Birth of Bio-politics: Lectures at the college de France 1978–79* (Basingstoke, UK: Palgrave Macmillan).
Franck, K. and Stevens, Q. (eds) (2007) *Loose Space: Possibility and diversity in urban life* (Oxford: Routledge).
Freund, P. and Martin, G. (2001) Moving bodies: Injury, disease and the social organisation of space, *Critical Public Health*, 11(3), pp. 203–214.
Furness, Z. (2005) *Put the Fun Between Your Legs! The politics and counterculture of the bicycle*, unpublished PhD thesis (University of Pittsburgh).
Furness, Z. (2007) Critical mass, urban space and velomobility, *Mobilities*, 2(2), pp. 299–319.
Galloway, A. (2004) Intimations of everyday life: Ubiquitous computing and the city, *Cultural Studies*, 18(2/3), pp. 384–408.
Gandy M, 2002 *Concrete and Clay: Reworking nature in New York City* (Cambridge, MA: MIT Press).

Golub, A., Hoffman, M. and Lugo, A. (eds) (2016) *Bicycle Justice and Urban Transformation: Biking for all?* (Abingdon, UK: Routledge).

Greene, D. and Wegener, M. (1997) Sustainable transport, *Journal of Transport Geography*, 5(3), pp. 177–190.

Guthman, J. (2009) Teaching the politics of obesity: Insights into neoliberal embodiment and contemporary biopolitics, *Antipode*, 41(5), pp. 1110–1133.

Hackworth, J. (2007) *The Neoliberal City: Governance, ideology and development in American urbanism* (Ithaca, NY: Cornell University Press).

Harvey, D. (1982) *The Limits to Capital*, 2nd edition (London: Verso).

Harvey, D. (1989) From managerialism to entrepreneurialism: The transformation in urban governance in late capitalism, *Geografiska Annaler*, 71B(1), pp. 3–17.

Harvey, D. (1991) *The Condition of Post-Modernity: An enquiry into the origins of cultural change* (Malden, MA: Wiley- Blackwell).

Harvey, D. (2003a) The fetish of technology: Causes and consequences, *Macalester International*, 13, pp. 3–30. Available at: http://digitalcommons.macalester.edu/cgi/view content.cgi?article=1411&context=macintl [accessed 30 November 2014].

Harvey, D. (2003b) *The New Imperialism* (Oxford: Oxford University Press).

Hengst, H. (2000) *Reconquering Urban Spots and Spaces? Children's public(ness) and the script of media industry*, Working Paper 18, Child and Youth Culture (Odense University: The Department of Contemporary Cultural Studies). Available at: hum.sdu.dk/center/kultur/arb_pap/ [accessed January 2007].

Hoffman, M. and Lugo, A. (2014) Who is 'world class'? Transportation justice and bicycle policy, *Urbanities*, 1, pp. 45–61.

Horton, D. (2006) Environmentalism and the bicycle, *Environmental Politics*, 15(1), pp. 41–58.

Horton, D. (2007) 'Fear of cycling', in Horton, D., Rosen, P. and Cox, P. (eds) *Cycling and Society* (Aldershot, UK/Burlington, VT: Ashgate), pp. 133–152.

Hummel, J. and Van der Duim, R. (2016) SNV's modes of ordering: Organizing tourism as development practice, *Tourism Management*, 57, pp. 312–322.

Jessop, B. (2002) *The Future of the Capitalist State* (Oxford: Polity Press/Malden, MA: Blackwell).

Jones, A. (2013) A tripartite conceptualisation of urban public space as a site for play: Evidence from South Bank, London, *Urban Geography*, 34(8), pp. 1144–1170.

Jones, A. (2016) Orchestrated public space: The curatorial dimensions of the transformation of London's Southbank Centre, *LSE Research Online* [accessed August 2019].

Jones, P. (2005) Performing the city: A body and a bicycle take on Birmingham, UK, *Social and Cultural Geography*, 6(6), pp. 813–830.

Jones, P. (2012) Sensory indiscipline and affect: A study of commuter cycling, *Social and Cultural Geography*, 13(6), pp. 645–658.

Jungnickel, K. and Aldred, R. (2014) Cycling's sensory strategies: How cyclists mediate their exposure to the urban environment, *Mobilities*, 9(2), pp. 238–255.

Junvaria, J. (2018) *Performing Cycling: Cultural dimensions of cycling practices*, unpublished MSc research thesis (University of Salford).

Kearns, R., Collins, D. and Neuwelt, P. (2003) The walking school bus: Extending children's geographies?, *Area*, 35(3), pp. 285–292.

Kidder, J. (2005) Style and action: A decoding of bike messenger symbols, *Journal of Contemporary Ethnography*, 34(2), pp. 344–367.

Kloof, A., Bastiaanssen, J. and Martens, M. (2014) Bicycle lessons, activity participation and empowerment, *Case Studies on Transport Policy*, 2, pp. 89–95.

Kristensen, K. (2013) *Michel Foucault on Bio-power and Biopolitics*, unpublished MSc thesis (University of Helsinki).

Krivy, M. (2018) Towards a critique of cybernetic urbanism: The smart city and the society of control, *Planning Theory*, 17(1), pp. 8–30.

Lam, T.F. (2018) Hackney: A cycling borough for whom?, *Applied Mobilities*, 3(2), pp. 115–132.

Latour, B. (2008) 'Where are the missing masses? The sociology of a few mundane artifacts', in Bijker, W.E. and Law, J. (eds) *Shaping Technology/Building Society: Studies in sociotechnical change* (Cambridge, MA: MIT Press), pp. 225–258.

Lattimer, J. and Munro, R. (2006) Driving the social, *Sociological Review*, 54(s1), pp. 32–53.

Law, J. (2001, June) *Ordering and Obduracy*. Lancaster: Centre for Science Studies, Lancaster University. Available at: www.comp.lancs.ac.uk/sociology/papers/Law-Ordering-and-Obduracy.pdf

Law, J. (2007) *Actor Network Theory and Material Semiotics*. Available at: www.heterogeneities.net/publications/Law2007ANTandMaterialSemiotics.pdf [accessed 30 June 2017].

Law, R. (1999) Beyond 'women and transport': Towards new geographies of gender and daily mobility, *Progress in Human Geography*, 23(4), pp. 567–588.

Lee, D.J. (2018) *Delivering Justice: Food delivery cyclists in New York City*. Available at: https://academicworks.cuny.edu/cgi/viewcontent.cgi?article=3854&context=gc_etds [accessed December 2019].

Lucas, K. (1998) Upwardly mobile: Regeneration and the quest for sustainable mobility in the Thames Gateway, *Journal of Transport Geography*, 6(3), pp. 211–225.

Lumsdon, L. (2000) Transport and tourism: Cycle tourism – a model for sustainable development?, *Journal of Sustainable Tourism*, 8(5), pp. 361–377.

Lumsdon, L. and Tolley, R. (2001) The national cycle strategy in the UK: To what extent have local authorities adopted its model strategy approach?, *Journal of Transport Geography*, 9, pp. 293–301.

Lupton, D. and Noble, G. (2002) Mine/not mine: Appropriating personal computers in the academic workplace, *Journal of Sociology*, 38(1), pp. 5–23.

Marx, K. (1990) *Capital, Volume 1*, trans. Ben Fowkes (London: Penguin Group).

Massey, D. (1991) Flexible sexism, *Environment and Planning D: Society and Space*, 9, pp. 31–57.

McCann, E. (2011) Urban policy mobilities and global circuits of knowledge: Toward a research agenda, *Annals of the Association of American Geographers*, 101(1), pp. 107–130.

McClintock, H. (1992) *The Bicycle and City Traffic* (London: Bellhaven).

McClintock, H. and Cleary, J. (1996) Cycle facilities and cyclists' safety, *Transport Policy*, 3(1–2), pp. 67–77.

McFall, L. (2009) Devices and desires: How useful is the 'new' new economic sociology for understanding market attachment, *Sociology Compass*, 3(2), pp. 267–282.

Meador, D. (2016) 'Cycling on the U.S./Mexican border with Doble Rueda bicycle collective in Matamoros, Tamaulipas', in Golub, A., Hoffman, M. and Lugo, A. (eds) *Bicycle Justice and Urban Transformation: Biking for all?* (London: Routledge), pp. 156–171.

Merriman, P. (2006) 'Mirror, signal, manoeuvre': Assembling and governing the motorway driver in late 1950s Britain, *Sociological Review*, 54(s1), pp. 75–92.

Miller, D. (ed.) (2001) *Car Cultures* (Oxford: Berg).
Millington, B. (2014) Smartphone apps and the mobile privatization of health and fitness, *Critical Studies in Media Communication*, 31(5), pp. 479–493.
Mould, O. (2018) *Against Creativity* (London: Verso).
Muniesa, F., Millo, Y. and Callon, M. (2007) An introduction to market devices, *Sociological Review*, 55(2), pp. 1–12.
Murdoch, J. (1997) Inhuman/nonhuman/human: Actor-network theory and the potential for a non-dualistic and symmetrical perspective on nature and society, *Environment and Planning D: Society and Space*, 15(6), pp. 731–756.
Newman, P., Kenworthy, J. and Vintila, P. (1995) Can we overcome automobile dependence? Physical planning in an age of cynicism, *Cities*, 12(1), pp. 53–65.
Nixon, D. (2012) A sense of momentum: Mobility practices and dis/embodied landscapes of energy use, *Environment and Planning A*, 44(7), pp. 1661–1678.
Palmer, C. (1996) *A Life of Its Own: The social construction of the Tour de France*, unpublished PhD thesis (University of Adelaide).
Palmer, C. (2001) Shit happens: The selling of risk in extreme sport, *Australian Journal of Anthropology*, 13(3), pp. 323–336.
Panter, J., Heinen, E., Mackett, R. and Ogilvie, D. (2016) Impact of new transport infrastructure on walking, cycling, and physical activity, *American Journal of Preventive Medicine*, 50(2), pp. e45–e53. doi:10.1016/j.amepre.2015.09.021.
Parkin, J. (2003) Comparisons of cycle use for the journey to work from the '81, '91 and 2001 censuses, *Traffic Engineering and Control*, 44(8), pp. 299–302.
Parkin, J., Ryley, T. and Jones, T. (2007) 'Barriers to cycling: An exploration of quantitative analyses', in Horton, D., Rosen, P. and Cox, P. (eds) *Cycling and Society* (Aldershot, UK: Ashgate), pp. 67–82.
Parkin, J., Wardman, M. and Page, M. (2008) Estimation of the determinants of bicycle mode share for the journey to work using census data, *Transportation*, 35(1), pp. 93–109. doi:http://dx.doi.org/10.1007/s11116-007-9137-5.
Pucher, J. and Buehler, R. (2011) *Analysis of Bicycling Trends and Policies in Large North American Cities: Lessons for New York*, University Transportation and Research Center, Region 2 Final Report. Available at: www.utrc2.org/publications/analysis-bicycling-trends-and-policies-large-north-american-cities-lessons-new-york [accessed 20 November 2014].
Pucher, J., Dill, J. and Handy, S. (2010) Infrastructure, programs, and policies to increase bicycling: An international review, *Preventative Medicine*, 50(Suppl 1), pp. S106–125. doi:10.1016/j.ypmed.2009.07.028.
Raje, F. (2007) The lived experience of transport structure: An exploration of transport's role in people's lives, *Mobilities*, 2(1), pp. 51–74.
Redshaw, S. (2007) Articulations of the car: The dominant articulations of racing and rally driving, *Mobilities*, 2(1), pp. 121–141.
Rietveld, P. and Daniel, V. (2004) Determinants of bicycle use: Do municipal policies matter? *Transportation Research Part A*, 38, pp. 531–550.
Shapiro, A. (2018) Between autonomy and control: Strategies of arbitrage in the 'on-demand' economy, *New Media and Society*, 20(8), pp. 2954–2971.
Sheller, M. (2004) Automotive emotions: Feeling the car, *Theory, Culture and Society*, 21(4–5), pp. 221–242.
Sheller, M. and Urry, J. (2006) The new mobilities paradigm, *Environment and Planning A*, 38, pp. 207–226.

Skinner, D. and Rosen, P. (2007) 'Hell is other cyclists: Rethinking transport and identity', in Horton, D., Rosen, P. and Cox, P. (eds) *Cycling and Society* (Aldershot, UK: Ashgate), pp. 83–96.
Song, Y., Preston, J. and Ogilvie, D. (2017) New walking and cycling infrastructure and modal shift in the UK: A quasi-experimental panel study, *Transportation Research Part A: Policy and Practice*, 95, pp. 320–333. doi:10.1016/j.tra.2016.11.017.
Southworth, M. (2005) Measuring the liveable city, *Built Environment*, 29(4), pp. 343–354.
Spinney, J. (2006) A place of sense: A kinaesthetic ethnography of cyclists on Mt Ventoux, *Environment and Planning D: Society and Space*, 24(5), pp. 709–732.
Spinney, J. (2007) 'Cycling the city: Non-place and the sensory construction of meaning in a mobile practice', in Horton, D., Rosen, P. and Cox, P. (eds) *Cycling and Society* (Aldershot, UK: Ashgate), pp. 25–46.
Spinney, J. (2008) *Cycling the City: Movement, meaning and practice*, unpublished PhD thesis (Royal Holloway University of London).
Spinney, J. (2010) 'Mobilising sustainability: Partnership working between a pro-cycling NGO and local government in London (UK)', in Fudge, S., Peters, M. and Jackson, T. (eds) *Low Carbon Communities: Imaginative approaches to combating climate change* (Cheltenham: Edward Elgar), pp. 89–107.
Spinney, J. (2016) Fixing mobility in the neoliberal city: Cycling policy and practice in London as a mode of political–economic and biopolitical governance, *Annals of the American Association of Geographers*, 106(2), pp. 450–458.
Spinney, J. and Lin, W. (2019) (Mobility) fixing the Taiwanese bicycle industry: The production and economisation of cycling culture in pursuit of accumulation, *Mobilities*, 14(4), pp. 524–544.
Spotswood, F., Chatterton, T., Tapp, A. and Williams, D. (2015) Analysing cycling as a social practice: An empirical grounding for behaviour change, *Transportation Research Part F: Traffic Psychology and Behaviour*, 29, pp. 22–33.
Stehlin, J. (2014) Regulating inclusion: Spatial form, social process, and the normalization of cycling practice in the USA, *Mobilities*, 9(1), pp. 21–41.
Steinbach, R., Green, J., Datta, J. and Edwards, P. (2011) Cycling and the city: A case study of how gendered, ethnic and class identities can shape healthy transport choices, *Social Science and Medicine*, 72(7), pp. 1123–1130.
Stevens, Q. (2007) *The Ludic City: Exploring the potential of public spaces* (Oxford: Routledge).
Stevens, Q. and Dovey, K. (2004) Appropriating the spectacle: Play and politics in a leisure landscape, *Journal of Urban Design*, 9(3), pp. 351–365.
Strömberg, M. (2017) Ride share, *Planning*, 83(4), pp. 26–27.
Thoburn, N. (2007) Patterns of production: Cultural studies after hegemony, *Theory, Culture and Society*, 74(3), pp. 79–94.
Tolley, R. (1994) Conference reports: Velo-City Conference 6–10 September 1993, Nottingham, UK, *Journal of Transport Geography*, 2(1), pp. 69–71.
Tolley, R. (1996) Green campuses: Cutting the environmental costs of commuting, *Journal of Transport Geography*, 4(3), pp. 213–217.
Torjesen, I. (2017) Cycling to work has substantial health benefits, study finds, *BMJ*, 357. doi:10.1136/bmj.j1944.
Tyfield, T. (2015) 'What is to be done?' Insights and blind spots from cultural political economy(s), *Journal of Critical Realism*, 14(5), pp. 530–548. doi:10.1179/14767430 15Z.00000000082.

Vaaranen, V. and Wieloch, N. (2001) Car crashes and dead end careers: Leisure pursuits of the Finnish subculture of the kortteliralli street racing, *YOUNG: Nordic Journal of Youth Research*, 10(1), pp. 42–58.

Van Der Duim, R. (2005) TOURISMSCAPES: An actor-network perspective, *Annals of Tourism Research*, 54(4), pp. 961–976.

Vandaele, K., Piasna, A. and Drahokoupil, J. (2019) *'Algorithm Breakers' are not a Different 'Species': Attitudes towards Trade Unions of Deliveroo Riders in Belgium*, European Trade Union Institute. Available at: www.etui.org/content/download/36345/364997/file/WP-2019-06-web.pdf

Vigar, G. (2002) *The Politics of Mobility: Transport, the environment and public policy* (London: Spon Press).

Webb, E. (2014) *Bus Passes, Active Travel and Health*, ICLS Occasional Paper, 11(3) (London: University College London).

Whitelegg, J. (1983) Road safety: Deceit, complicity and the bankruptcy of science, *Accident Analysis and Prevention*, 15(2), pp. 153–160.

Whitelegg, J. (1987) A geography of road traffic accidents, *Transactions of the Institute of British Geographers*, 12, pp. 161–176.

Williamson, B. (2015) Algorithmic skin: Health-tracking technologies, personal analytics and the biopedagogies of digitized health and physical education, *Sport, Education and Society*, 20(1), pp. 133–151.

Wood, A. (2014) The politics of policy circulation: Unpacking the relationship between South African and South American cities in the adoption of bus rapid transit, *Antipode*, doi:10.1111/anti.12135.

Woodcock, J. Bannister, D., Edwards, P. and Prentice, A. (2007) Energy and transport, *The Lancet*, 370(9592), pp. 1078–1088.

Xie, L. and Spinney, J. (2018) 'I won't cycle on a route like this; I don't think I fully understood what isolation meant': A critical evaluation of the safety principles in Cycling Level of Service (CLoS) tools from a gender perspective, *Travel Behaviour and Society*, 13, pp. 197–213.

Yang, L., Sahlqvist, S., McMinn, A., Griffin, S.J. and Ogilvie, D. (2010) Interventions to promote cycling: Systematic review, *BMJ*, 341, pp. 1–10. doi:10.1136/bmj.c5293.

Zuboff, S. (2019) *Surveillance Capitalism: The fight for a human future at the new frontier of power* (New York: Hachette Book Group).

Zuev, D., Tyfield, D. and Urry, J. (2018) Where is the politics? E-bike mobility in urban China and civilizational government, *Environmental Innovations and Societal Transition*. doi:10.1016/j.eist.2018.07.002.

3 Making up the (productive) cycling subject

Excluding the 'non-standard' user in cycle infrastructure design

Introduction

In the previous chapter I selectively reviewed research on cycling. In doing so I established that whilst there has been great progression, earlier accounts of cycling tended to homogenise the cyclist as a subject. Concerned with recuperating cycling from low levels, a lot of research has assumed that 'if we build it, they will come'. However, the 'they' in this mantra has until recently not been problematised. In particular, social and cultural differences that might preclude users from taking up cycling have been glossed over. Accordingly, this chapter explores who is and isn't part of the cycling 'boom', arguing that women, children, older people and those with impairments are still under-represented in cycling as a practice in the UK. This contrasts to some extent with more established cycling nations, where cities like Amsterdam and Copenhagen have more proportional levels of women cycling, although they too are yet to achieve equity with regard to older people, children, those with impairments and ethnic minorities.

Though the reasons for these absences are numerous and complex, this chapter emphasises the role of design as a form of 'ordering' that serves to manage and control the appropriate range of performances and users that are materialised and legitimised as cycling subjects. Here I argue that narrow conceptions of cycling as an individual instrumental practice more or less implicitly inform design guidance and planning practice to the extent that where cycle infrastructure exists, it tends to reify this narrow definition of cycling and exclude alternative uses and users. As a result, the orderings of cycle design and those that result from cycle design (such as material environments) privilege and bring into being an adult, able-bodied male commuting cyclist at the expense of other subjectivities. Such a position is untenable in the context of sustainability because, as Koglin (2011) has stated, 'developing bike-friendly policies and providing cyclists with good infrastructure are matters of equity in the city. The city is for everybody and all have the right to be in the city (226).

After introducing the importance of design in the context of actor network theory (ANT) and mobilities as a framework to understand the production of distinct subjectivities, I then go on to discuss processes of standardisation in transport design more broadly. In doing so I show how design techniques of counting and modelling have elided and excluded the affordances of cyclists, making the cycling subject de facto less likely to emerge. I then go on to focus more explicitly on the design of cycle-specific infrastructure using as exemplars two subjects more likely to be excluded from cycling: children and the physically impaired. With regard to children, I show how current cycle design guidance fails to account for their valuations of cycling, ontologically excluding them from performing cycling in ways that whilst 'unproductive' are more reflective of their desires and identities. Using the case of physical impairments I then go on to show how design guidance creates a distinction between those able to cycle and those unable through a failure to accommodate the different ways in which impaired bodies hybridise with 'non-standard' cycles and infrastructure. Ultimately I argue that if we are to bring into being a wider range of cycling subjects, we need to reorder cycling, and much of that starts with the assumptions we import into design regarding the range of user performances and desires. In doing so, this chapter also speaks to the book's broader argument regarding the exclusion of 'unproductive' subjects in contemporary manifestations of cycling. It is clear that children and the impaired are more or less explicitly excluded from cycling because the variants of cycling that they might perform would be too slow or troublesome to accommodate in a system that prioritises speed and efficiency above any other qualities.

Who are the absent users in cycling?

Whilst little has changed at the UK national level, with cycling stuck at around 2% mode share and 1% of all distance travelled (DfT 2017: 5), some towns and cities have seen significant increases in the modal share of cycling, particularly when it comes to the journey to work (Aldred *et al.* 2016: 37). Whilst we might expect to find that overall increases in cycling have been accompanied by an increase in the proportion of women and minority groups cycling, as is the case in many places with high cycling mode share, this has not been the case in the UK. Aldred *et al.* (2016), in their analysis of 2001/2011 UK census data, found that despite overall increases in cycling in many towns and cities, the ratio of women and elderly cyclists had stayed constant or even declined. DfT statistics from 2017 confirm, with men making three times as many trips as women (2017: 5). The significance is that any overall growth in cycling numbers hides the fact that many social groups are still under-represented (Aldred *et al.* 2016; TfL 2010).

Female participation in cycling is viewed as an 'indicator' of a cycling-friendly culture and environment (Garrard *et al.* 2006: 11). However, it is widely noted that men heavily outnumber women in the proportion of cyclists (Krizek *et al.* 2005; Moudon *et al.* 2005; Emond *et al.* 2009;

Winters *et al.* 2007). Whilst there are a number of reasons for this, perhaps the most important is that traditional social and domestic gendered divisions of labour mean women are more likely to fulfil social reproductive roles such as child-caring and shopping (Emond *et al.* 2009; Garrard *et al.* 2012). However, because standard bicycle designs do not allow children or substantial cargo to be carried, and because women's journeys tend to be more complex, women opt to use the car or public transport in order to fulfil these roles. As Garrard *et al.* (2012) comment, using examples from Copenhagen and Amsterdam, given the correct infrastructural environments,[1] it is clearly possible to create conditions where non-standard cycles can be used to fulfil escort and shopping trips.

Cycling in the UK is less common than walking among the older population, accounting for only 1% of all journeys made by people aged 65 and older in the UK compared with 23% in the Netherlands, 15% in Denmark and 9% in Germany (Pucher and Buehler 2011). Whilst the reasons for this decline are again complex, a recent UK study of cycling in later life noted in particular that many participants sought to adapt their bicycles and some wished to use a trike so that they could continue to cycle despite a decline in physical mobility. However, those participants in the study who had opted to use a trike found that it was difficult or impossible to negotiate both cycle-specific and highways infrastructure with a non-standard or adapted cycle (Jones *et al.* 2016).

If adult cyclists in the UK are an endangered species, children are all but extinct. The National Travel Survey (2019) found that in 2017 children aged up to 16 cycled on average 23 miles per year in around 13 journeys, less than half that of the average 55-year-old. The figures for journeys to school are even worse, with only 1% of primary school children in the UK cycling to school, down from 3% in 2013; just 3% and 1% of secondary school children cycle in England and Wales, respectively (www.cyclinguk.org/statistics). These figures are in line with broader trends over the last 50 years of rapid decline in children's independent mobility (Barker 2008).

Mobility impairment is the most commonly reported form of disability in the UK, with 57% of disabled people experiencing this (DWP 2014). However, as Clayton *et al.* (2017) note, most mobility impairments are 'invisible', with the result that mobility impairment, and the barriers people face in relation to this, are under-appreciated in transport generally, and thus under-represented in policy and design (Clayton *et al.* 2017: 453). Clayton *et al.* (2017) go on to state that whilst:

> the ability to cycle with a disability may not be in question, there remain issues however about – for example – mounting and dismounting, walking alongside the cycle, balancing at very low speed or while static, and the ability to manoeuvre and successfully park the cycle.
>
> (Clayton *et al.* 2017: 454)

Their study of disabled cycle users emphasises the point that despite some mention of it in current cycle design guidance, 'the issue of cycle dimensions remains an issue, and will require remedial action to overcome deficiencies in practice' (Clayton *et al.* 2017: 461). They go on to note that whilst:

> poor and inadequate infrastructure is a frustration … for all cyclists, for disabled cyclists often this makes infrastructure wholly inaccessible, and thus is a directly limiting factor in disabled people's ability to maintain an active lifestyle, and to receive the health and wellbeing benefits of doing so.
>
> (ibid.)

As Aldred *et al.* (2016) conclude, 'commuter cycling in the UK is concentrated among specific demographic groups, with regular commuters more likely to be male, white, able-bodied, young adults' (29). As this suggests, whilst there are complex reasons why cycling levels for women, the elderly, disabled and children remain low, *a cross-cutting factor is that they have different goals or affordances from other cyclists, and are more likely to require a non-standard cycle, and corresponding infrastructure that can accommodate these*. Whilst good progress is being made in some towns and cities on the implementation of separated infrastructure, it tends to favour standard cycles and cyclists over non-standard cycles. One of the key questions posed by Schliwa *et al.* (2015) is whether existing cycle infrastructure can cope with non-standard bikes and trikes often necessary for older people, the disabled, families (not to mention freight) by bike? In particular, Schliwa *et al.* (2015) raise the need to revisit cycle lane widths, turning circles and parking for non-standard cycles in order to enable a broader range of users to take up cycling.

Ordering cycling: design and materiality

Evidently cycle design has a key role to play in the exclusion of particular users. In this chapter, I wish to try and understand how these distinctions emerge within design as a form of ordering. Van Der Duim *et al.* (2013) state that ANT approaches to social science are less concerned with 'why' questions, and more with 'how' social life comes to take particular forms (5). Accordingly, I argue here that the current exclusion of particular users is a product of a particular ordering that brings into being a distinction between not only cyclists and drivers, but also between those who can cycle and those who cannot. Certainly if we cast the absence of particular users in cycling as a matter of concern rather than a matter of fact, we start to ask how such a state of affairs might come into being: what particular orderings produce and stabilise such an outcome if it is not inevitable?

My focus here will be on the role of design as a form of ordering because it attempts to control and manage cycling as a phenomenon (Hunt and

Wickham 1994, in Franklin 2004: 285) with regards to what constitute ontological priorities, their materialisation, and the range of users who can participate. The significance of this is that in such an approach, any distinction between users who are and aren't accommodated is seen not to be the cause of such orderings (e.g. a basis that informs design guidance), but rather these distinctions are seen as an effect of orderings (such as design guidance) and the ways in which they conceptualise and treat users (Law 1999).

In taking such an approach, orderings are seen to stem from narratives about the world and therefore represent attempts to materialise 'how things ought to be' (Franklin 2004). In line with this, a key argument I make is that the ways in which distinctions between cyclists are made within the professions of design (in particular transport planning and engineering) are premised upon attempting to bring the economically productive cyclist into being whilst marginalising other cycling subjects whose desires and affordances do not conform to such a narrative. Accordingly, of particular interest are the ways in which seemingly 'black boxed' networks and phenomena come into being and are stabilised in particular ways in order to exert political and economic control 'at a distance' (Van Der Duim *et al.* 2013: 5). It is in this context that I want to introduce design and design tools as a primary form of political control through which the narrative of the economically productive cyclist is produced and stabilised 'at a distance'.

Designing the mobile subject

Design is both verb and noun – process and product (Lawson 2006: 3) – something that we both do and experience as an outcome. Merriman has recently argued for a greater emphasis on design in mobilities research, stating that there is much more to be written on 'examining the history of the design, planning, engineering, labouring, consumption and use' of mobility spaces and artefacts (Merriman 2015: 88). There is of course a substantial body of work dealing with the relations between bodies, environments and designed artefacts, particularly from science and technology studies scholars and technological anthropologists (Akrich 1992; Fallan 2008; Ingold 2000; Michaels 2000a, 2000b; Oudshoorn *et al.* 2004, 2016). Where geographers have considered the role of design in relation to mobility, it has tended to focus on the artefacts produced and their (inter) mediating role between body and environment (Borden 2001; Merriman 2006; Spinney 2006; Spinney *et al.* 2017), or been treated as a more or less taken for granted part of the hybrid (Imrie 2000, 2001; Pyer and Tucker 2017; Stevens 2004). However, mobilities scholarship actively encourages us to view design as a force assembling societies and subjectivities in particular ways.

One of the first mobility objects that Urry (2000) turned his attention to in order to demonstrate this was that of the car. Urry found it baffling that the car had not received more sociological attention, positioned as it was as a 'neutral technology, permitting social patterns of life that would have more or less occurred anyway' (2000: 59). Sociology, it is argued, has ignored the ways in which the car has transformed the time-space of the urban dweller in particular: 'failing to consider how the car reconfigures urban life, with novel ways of dwelling, travelling, and socialising in, and through, an automobilised time-space' (Sheller and Urry 2006: 210). For Urry, the car's significance as a designed object was that, through distinct ways of dwelling, socialising and travelling, it reconfigured civil society and subjectivity (ibid.). As Dant (2004) points out, the 'hybrid' driver-car 'is not a species resulting from chance mating but a product of human design, manufacture and choice' (62). Accordingly, Yaneva (2012) argues that 'we cannot understand how a society works without appreciating how design shapes, facilitates, conditions and makes possible everyday sociality' (280).

To understand design processes is then of fundamental importance because of the ways in which they act to afford or limit the potential for transformation and alternative visions of the world. As Akrich (1992) argues:

> technical objects have political strength. They may change social relations, but they also stabilise, naturalise, depoliticise and translate these into other media. After the event, the processes involved in building up technical objects are concealed … . There was, or so it seems, never any possibility that it could have been otherwise.
>
> (Akrich 1992: 222)

Consequently, design must be considered a process through which human intentionality is both given and hidden in objects, producing them as moral and political actants. Accordingly, where possible, we need to explore the design process to discover how artefacts come into being as much as we look at the impacts of design.

It is with this understanding in mind that Urry asks us to consider how, 'objects are materially produced and symbolically conceived of' (2000: 66). In the language of actor network theory (ANT), rather than 'black box' design processes, allowing products to be seen as singular and linear entities, Urry encourages us to unpack design and focus on the movements, contingencies and multiplicities that bring objects into being in particular configurations. From such a perspective, artefacts are assembled at disparate locations from numerous parts and must be understood as a distributed process: the intersection of forms of expertise and values (Yaneva 2009). Design is thus an inherently political practice, even if design professionals have been at pains to obscure this with recourse to aesthetics and functionalism. Designed artefacts are the materialisation – and indeed fixing and representation – of particular values, interests and dispositions. As such they tell a particular story, or more accurately they give us the 'ending' – the outcome. As Yaneva states, 'together with legal, technical, artistic and religious ties, design helps make the social durable' (Yaneva 2012: 281). However, we cannot tell just by looking at something how it has come to take this shape and not that or why it enables this use but disables that use.

Yaneva argues that to tackle design as a type of connector, 'as a mechanism for setting the world in motion', we need to go beyond the discourses of designers and inventors (2012: 282). We need to also attend to the cultures and practices of designers, 'to follow what designers and users do in their daily and routine actions' (ibid.). Storni (2012) suggests a focus on 'the process of design, in the becoming of its results and in the relation between process and outcome' to foreground the movements and transformations between product and process. One way of doing this is to focus on the interplay between virtualised and 'invisible' aspects of design such as standards, regulations and methodologies to understand the ways in which they bring particular affordances into being. An approach which incorporates these offers us the possibility of making visible the agency of many actors and actants not visible in traditional design thinking: the importance of various non-humans comes to the fore (Yaneva 2012: 283).

It is here that I want to introduce planners and engineers as the product designers of the public realm. A central role of these professionals is to design mobility spaces that afford particular forms of movement and enable citizens to carry out their daily lives in ways desirable by the government of the day (Sennett 1994). Thus the process of infrastructure design is one of aligning user preferences with those of the state: it is, as Foucault has stated, a key element enabling the conduct of conduct. However, like much product design, the conceptualisation of the user in transport planning is often partial at best, with the result that some users may be excluded from using the product of the built environment because either they are physically incapable of using it as designed, or the script it embodies cannot be entangled in their daily lives. Congruently, as Oudshoorn *et al.* (2004) state, in order to 'understand how technological artifacts come to incorporate barriers against specific groups of users, we need to shift the analysis to users in the semiotic sense: users as imagined by the designers of a technology' (31).

Conceptualising the cyclist in infrastructure design

Having set up a theoretical premise for exploring the imagined user within transport infrastructure, in this section I discuss not only the ways in which cyclists as a group with particular affordances have been written out of transport planning, but also how particular social groups have been precluded from participating in cycling. In this first section I argue that cyclists in general have been written out of planning and design due to an emphasis on predictable standardised vehicular movements to which they do not conform. I argue here that this is operationalised through the tools and methods of design, including modelling and design guidance. These tools have historically been poor at conceptualising users other than automobile users and as a result have produced a particular transport ordering that marginalises cycling and makes it less likely that the cycling subject will emerge.

Under modernism the streets became a place for getting from A to B: a system of transport rather than a place to live in (Fyfe 1998: 1). Streets that had once been for walking and socialising were now for moving, and with the emphasis of the modernists on the city as machine, it was an increasingly mechanical movement with the street reconstructed as a machine for traffic, 'used exclusively by fast moving mechanical vehicles, and free from pedestrians and building fronts'. For Corbusier – the grandfather of modernism – a street made for speed was a street made for (economic) success and the sacrifice of many traditional leisurely, social and playful street activities was a price worth paying (Corbusier, quoted in Fyfe 1998: 2–3).

It is within the context of this desire for fast and ordered movement that automobility grew at rapid rates in the UK. In 1903 cars were licensed and given the full freedom of English roads. However, it was not until after the First World War that motoring really took off (see, for example, Morton 1933). By 1920, there were 187,000 private cars and 88,000 public vehicles on the UK's roads (Fullerton 1975: 30). After 1920 the growth in vehicles accelerated and outstripped the capacity of the existing roads, reaching 1 million by 1930 and 13 million by 1972 (Fullerton 1975: 33).

The increasing complexity and dangers posed by automobility consequently brought a new direction to planning and civic improvements based upon the 'use of expert knowledge, state regulatory mechanisms, and public welfare provision' (Southworth and Ben-Joseph 2003: 66). The first Traffic Engineering Handbook was published in 1942 as the basis for professional practice and it was not long before standardisation became the mantra (Southworth and Ben-Joseph 2003: 83). The process of standardisation attempted to conceive a system based upon the affordances of motorised vehicles. The new Taylorist measures, as Cresswell (2006) notes, were geared towards 'the transformation of learned habit – embodied habitual mobility – into a rigorous and scientifically coded abstraction of human motion' (86). This transformation of embodied mobility was based upon the 'elimination of superfluous and awkward motions' (ibid.).

Murdoch (1997) defines such standardisation as a spatial, temporal or spatio-temporal segmentation of the world. According to Murdoch, such standardisations are more than just discursive constructs; they are built into and embedded in every feature of the built environment and thus act to enclose, confine and organise behaviour in particular spaces (365). The results of such standardisations are that only particular affordances, for example, those of motorised transport or standard cycle users, are included in design guidance, planning tools and ultimately materialised in the built environment.

Designing out cycling – model movements

As I discuss in Chapter 4, there has been some 'progress' in the last 15 years, with the UK (notably TfL in London) starting to adopt and adapt Dutch standards of cycle infrastructure. However, as I elaborate upon here, not only do these fail to accommodate any diversity within the category of cyclist, they are not always accompanied by tools and modelling that enable designers to deliver progressive infrastructure. Instead, the focus remains on standard and predictable (vehicular) movements as part of the process of standardisation.

Professional planning in its modern form relies on designing the city using 'paper' and models for subsequent delivery to clients (Gehl 2001: 41). These tools have generally been of a meta-scale and technical in nature – large data sets and computer models used to analyse, predict and provide (Bannister 2002: 17). This point was reinforced in interviews with transport professionals who cited the key tools used to understand the movements of cyclists as traffic counts and observations, computer modelling, computer-aided design (CAD) and 'best practice' guidelines, particularly the *Cycle Design Standards* (either LCDS in London/England or WCDG in Wales), *Manual for Streets*, and the Dutch *CROW*.

The first step towards standardising movement is to reduce cyclists (and indeed all road users) to flows and counts: 'the assessment process requires current flow data for pedestrians, cyclists, other users … and motor vehicles. Surveys are needed for both the route being considered and any alternative routes' (Local Transport Note 2/04: 11). The use of such tools was also raised in interviews with transport professionals, where the emphasis was on counts and eliciting the views of campaigners (Planner 2, 23 June 2005).

Through an emphasis on statistical and quantifiable apprehension of behaviour on the roads, the individual disappears and becomes instead the 'dividual', 'by means of forms of calculation executed on bodies of population. The calculation is based on an ethics of personal renunciation and de-personalisation, transforming the individual to a number, a dividend or a "dividual"' (Forstorp 2006: 97). Using such numerical methods, all forms of road user can be seen to share the same characteristics: the unique and located individual is erased and replaced by the generalised and standardised anonymity of numbers. As Forstorp goes on to comment, the power of such methods is that, 'the transparency and self-evidential character of numbers are often taken for granted as an indisputable matter of fact' (Forstorp 2006: 98).

Gathering data on cyclists, as suggested here, takes the form of traffic counts and provides an idea of, for example, the number of cyclists per hour or per light change. Such an approach reduces the act of cycling to a flow in line with other vehicle flows in an attempt to let engineers calculate what they have to do to achieve maximum flow and efficiency with the available space and time. In the process of designing for cycling, once data on the numbers and types of vehicle is gathered, this is used to model the possible layouts and timings of junctions. However, as planners and engineers pointed out in interviews, these models predicting movement were largely concerned with the movements of motor vehicles:

> Probably the motor car is the driver as it were for getting highway schemes and traffic schemes done – that's what computer programs have been designed to do and deal with, and in terms of them looking at things in a mechanistic way, which is what engineers are primarily trained to deal with.
>
> (Engineer 5, 22 February 2005)

> We use a number of things to research what's needed. For example, vehicle swept paths (VSPs) so that we can see the path of the vehicle so you can choose various types of vehicle and put them through a junction on a plan and see what clearance spaces it needs to get through.
>
> (Engineer 3, 1 July 2005)

These reductive understandings of movement position cycling in specific ways. First, they presume that the behaviour of the cyclist (and indeed all vehicles) can be reduced to a series of two-dimensional plans and statistical predictions. Critics, such as Bannister (2002), have argued that such a positivistic approach straitjackets behaviour without attempting to understand it: 'as such it is the definitional and measurement issues which become important rather than understanding the reasons why people have to travel' (132). Certainly, as Bannister goes on to caution, the assumption of predictability trivialises actual behavioural decisions because it assumes that individuals are 'rational' decision-makers with full knowledge of what will happen during the trip they are about to embark upon (133). Transport professionals are aware of these shortcomings better than most, with a number of planners and engineers noting that the software and models were currently only capable of representing idealised rather than actual movements of pedestrians and cyclists:

> Well [those techniques are] common right the way across engineering. If you take a drawing in engineering they draw the road but none of the behaviour of road users. So the word assumption is quite appropriate but also there's a lot of direction of people to behave in a certain way. If you can't predict, then you try and direct.
>
> (Planner 1, 24 June 2005)

It is clear from their comments that planners and engineers understand that the complexities of mobility are greater than the reality represented in their computer models and drawings. However, there was also a feeling of pragmatism, that this was something that they could do little about and that alongside the views of other stakeholders the models were all that they had to work with. This implies the agency of modelling software to reproduce distinctions, and indeed to try and produce a version of the cyclist that behaved in ways desired by the system designers, rather than the system reflect how the cyclist might want to move. In the same way that Law recounts how his projector asymmetrically shapes the social relations between himself and students when lecturing (1992: 382), modelling and CAD software shapes relations between designers and cyclists because it black-boxes a set of assumptions around cyclist performance that make them very difficult to change even if they are questioned: the planner and engineer in this scenario have little agency to change the prescribed performance of the cyclist unless a concerted effort is mounted to work with the software designers to change the assumptions.

Cycle campaigners and activists involved in the design process tended to have much stronger views on the shortcomings of such approaches and the need to resist them:

> What they [planners and engineers] do is they're used to drawing plans on maps and not looking at how cyclists and other people interact with the space. It's a bit like when there's a grassed area and they're going to put paths, what they do is they let people walk the area to see where it gets worn and then put the path down, and what they should do is realise where cyclists are going to go and make that. The planners are all people who can use Autocad or whatever and do traffic counts, but they don't just watch what happens or cycle on it themselves.
>
> (Campaigner 3, 16 February 2005)

> Road design is an abstract thing and when you look at diagrams of junctions and stuff and flows and desire lines, it's all very pure and beautiful but in no way can it or does it represent the randomness or the total need to be aware and the general chaos, and it is chaos; the reversing lorries, the u-turning taxis, the muppet cyclist with the shopping wrapped round the front wheel with headphones on.
>
> (Campaigner 5, 9 August 2005)

There is evidently some dissatisfaction with the modelling process and its inability to encompass the full performance of cycling. The need to reduce people's mobility to fit into models and diagrams means that people's mobility must appear to be predictable; that is to say, there must be a limited number of desirable behaviours displayed which can all be captured within the format of the models. Such reduction denies the bodily performance of diverse mobilities and replaces it with an idealised and fictional appearance of vehicular mobility defined in documents like *The Highway Code*. As Imrie (2000) states, current 'wisdom' seeks to accommodate and indeed materialise predictable and productive mobility rather than conceiving of mobility as a messy, unpredictable and dynamic reality (1644).

The results in relation to the design of transport spaces have been that only particular affordances – those of motorised transport – have been materialised in the built environment. Perhaps most notably, there has been a failure to take into account any kinaesthetic dimensions of movement in particular, as Bendixson notes: 'the first consideration has been to prevent pedestrians from getting in the way of vehicles (the safety of bicyclists has been ignored). The psychological and physical requirements of people on foot have been disregarded' (Bendixson 1974: 57).

In the UK, the 1960s, for example, saw the wholesale introduction of one-way schemes to expand the capacity of streets (Buchanan *et al.* 1980: 266). The results of such schemes were reported, as hoped, as a reduction in accidents and reduced delays, but over increased distances: 'no allowance was made for pedestrian time or environmental dis-benefits which could have reduced the apparent worth of the schemes' (268–269). The main reason that no allowances were made for cyclists in such schemes was because, due to the reductive tools and assumptions already described, cyclists were assumed to have the same affordances and requirements as their motorised counterparts. Consequently, as Bendixson goes on to lament, a 1970s report into transport was:

> pessimistic about the future of cycling and says nothing about the inhibiting effect on it of heavy traffic, lengthy one way detours and traffic signals set to suit thirty mile an hour cars and not cyclists quietly pedalling along at half that speed.
>
> (Bendixson 1974: 85)

Whilst such a situation disadvantages both pedestrians and cyclists, I argue here that it disadvantages cyclists more because they are deemed to be vehicles and therefore must operate within a system designed overwhelmingly to accommodate motorised vehicles. Whilst pedestrians must intermittently integrate with this system, they are not obliged to abide by vehicular rules constantly and can usually act as 'natural pythagoreans' and follow a desired line perceived to require the least expenditure of energy. The cyclist, on the other hand – as a vehicle – has increasingly had to take on the identity of the car in order to move around. The point that I want to draw out of this is that this became the case even though the rules governing the use of new infrastructure were geared towards a vehicle which was vastly different from the bicycle in size, weight, propulsion, danger and concomitant other affordances. The new 'system' of automobility and the tools it relies on have erased the body of the person in favour of the body of the car, instituting a new set of system rules that all vehicles must abide by even though they are essential only for the rapid and 'safe' transit of motorised vehicles. The result of this is environments that actively militate against cycling because they fail to accommodate the less predictable, more fluid, more playful and human-powered movements that the cyclist is capable of. Thus the standardisations inherent in planning processes that privilege movements based upon motor vehicles actively discourage cycling as a practice. This is reflected in the historically low mode share of cycling in most UK towns and cities.

Cycle design guidance – non-standard users

Having set out how cycling in general has been marginalised to date through the design tools of planning and engineering, I now want to look more closely at how particular social groups, particularly children and those with disabilities, have been marginalised from participating in cycling as a practice because of their absence in the design process.

It is worth noting at the outset that there have been some recent developments in the past ten years that have resulted in some cyclist movements being modelled. For example, discussion with cycle engineers at a large UK charity tasked with delivering much of TfL's infrastructure in London showed that their software allowed them to model the widths and turning circles of standard cycles (Sustrans interview, August 2018). However, what also came to light was the fact that this software could only cope with standard upright safety bicycles: if the engineer wanted to design for non-standard cycles (cargo bikes, trikes, e-bikes, hand cycles, etc.), the software had no data upon which to model or develop the required infrastructure (ibid.). This, however, is not fundamentally a software problem: the models and processes of the planning and engineering professions are implicitly constrained by the evidence base and documents upon which notions of standardised movement are based. In the case of cycling, these include, but are not limited to, manuals on engineering for cycling such as the *London Cycle Design Standards* (LCDS) (or *Welsh Active Travel Design Guidance* – ATDG – in Wales), *Manual for Streets* and Dutch CROW. If these documents do not adequately account for 'non-standard' users, it is extremely unlikely that the models that the guidance informs will. Accordingly, in the following sections I discuss the role of design guidance in shaping what kinds of cycle infrastructure are designed and built, and how this specifically limits some users from participating and indeed produces them as 'non-standard'.

Children in cycle design guidance

In this section I review two of the main/most up-to-date sources of cycle design guidance available in the UK: the CROW *Design Manual for Bicycle Traffic* (2007) and Parkin's *Designing for Cycle Traffic* (2018). I do not review the London Cycle Design Standards because children are not even mentioned in the guidance, which itself speaks volumes about the status of children in current cycle discourse. The purpose of this is to understand the extent to which children feature, and where included, how their needs are understood, what they are considered to value when cycling, and the extent to which this is accommodated in design principles.

The Dutch CROW guidance is deemed by many cycle planners and engineers to be the gold standard of design guidance given the recent history and mode share of cycling in many Dutch cities. With regard to its understanding of the potential cycling population, CROW (2007) states that the 'cycling population is characterised by a broad composition in terms of age, gender, physical capacities and reasons for cycling' (26). It then goes on to outline how certain capabilities (and corresponding ages) might be prioritised in different designs (ibid.). Following this, it cites differences in terms of physical capacity, defining this in terms of maximal 'muscular work' over a given time (ibid.); and mental capacity, the limits of which it argues will be reached much sooner by the young and elderly (27). What is evident is that the young (and old) are seen to deviate from a norm of competency that is defined by awareness, speed and control. The 'normal' and desirable cyclist, then, is one who is fast, aware and fully competent. Such delineations say nothing about any other differences between children and adults in terms of what they value in their journeys, rather viewing children as 'deficient adults'. The resulting design guidelines focus on instrumental requirements, notably cohesion, directness, comfort and safety, with the only non-instrumental requirement being the need for attractiveness (CROW 2007: 31–32), which the guidance later suggests will be accomplished through varied surroundings and good lighting (64).

Where CROW talks about recreational cycling, it distinguishes between cycle touring and mountain biking but refers mainly to the needs of touring cyclists when discussing recreation (85). Two things are notable here: first, that recreation is seen as distinct from instrumental commuting cycling and should be accommodated separately. The emphasis on separating more leisurely and playful forms of mobility from fast and 'productive' mobility mirrors Le Corbusier's subordination and separation of 'unproductive' activity. Second, these are evidently 'adult' ideas of recreational cycling with a corresponding design emphasis on ensuring visual attractiveness with regard to 'idyllic country roads' (91) and comfort (92). There is some mention of other forms of recreational movement in the form of skate boarders (140) and how they might be accommodated to 'reduce the risk of mutual nuisance' (141). However, these alternative small-wheeled modes of movement are not considered to be a legitimate form of mobility. To summarise, there is almost no mention of children in the CROW, and where they do feature, they are assumed to value the same qualities as adults (directness, attractiveness) but are pictured as being less capable of attaining them and thus in need of special treatment to allow for any 'unpredictable' behaviour.

There is no sense, however, that any differences identified between different users are translated into specific design guidance. CROW features a number of non-standard bicycles, including a child's bike, and discusses variables including speed, braking and acceleration. It even discusses the width 'envelope' of different cycles/cyclists and suggests that younger cyclists require more room due to more 'zig-zagging', side-to-side movement (47). Despite this, the resulting design parameters stipulate minimum cycle lane width values to accommodate zig-zagging based on adults moving at 20 km/h (48) with no design tolerances given for any other users who may be deemed unable to operate within these parameters.

One of the most recent updates to cycle design guidance in the UK comes from respected academic and engineer John Parkin's insightful and rigorous new volume *Designing for Cycle Traffic* (2018). With regard to user groups, *Designing for Cycle Traffic* states that design should be 'inclusive of anyone who wishes to cycle', citing the liveable cities mantra of accommodating a wide range of ages from '8–80' (2018: 36). Good design for all is couched as being that which is 'safely rideable on all types of cycles' and attractive and comfortable for all types of rider (37). Children are described as being 'keen to explore the world', with the discussion around what is required to enable children to cycle is around bikeability training and providing children with the skills to cope in unwelcoming environments (Parkin 2018: 37), e.g. to transform them into the kinds of cyclists imagined and desired by the system designers. Other than this, there is discussion of the need for cycle parking at schools to encourage children to cycle these journeys (164). In common with the CROW, despite some recognition of differences between children and adult cyclists with regard to what they might value and are capable of, this is not translated into design parameters or guidance, which in line with the CROW continue to reflect those of adult (male, able-bodied) 'standard' users.

One document that seeks to go beyond such adult understandings is the (non-cycle-specific) design guidance *Routes to Play* (2009) published by sustainable transport charity Sustrans. The document – which aims to increase awareness of children's active travel needs – clearly acknowledges the unstructured nature of children's play and the fact that children see no distinction between active travel and play, noting that 'walking and cycling provide excellent play experiences in themselves as well as allowing children and young people to get to play spaces and recreational facilities under their own steam' (Sustrans 2009: 9). However, despite acknowledgement that 'more child-friendly environments and children can be further encouraged by installing play features along the routes they are likely to use' (2009: 18), the document goes on to outline an explicit strategy for creating a 'play friendly built environment' with guidance drawn from the *Manual for Streets* on how to improve walking and cycling routes to play spaces that mainly centres on restricting traffic, reducing vehicle speeds, adequate signage and lighting (Sustrans 2009: 11, 15). Despite acknowledging the potential playfulness of mobility, the multiple forms it might take and that 'a play space reaches beyond its boundary', the report reverts to a focus on mobility that emphasises its function and a small number of potential modes:

'surrounding street environment should be assessed to identify barriers to walking, cycling, using a mobility aid or taking public transport' (Sustrans 2009: 15).

This functional separation of play and mobility spaces is also evident in a report by urban designers Lee and Menzies (2016) when discussing cycle infrastructure in Amsterdam in relation to play spaces. They note a belief that mobility places in the city are integrated with play spaces, yet they go on to relay the fact that they 'discovered a great number of playgrounds, sports facilities, playable spaces, artworks and natural settings connected to homes, schools and places of work by this dense network of walkable, cycleable routes'. Whilst evidently a step in the right direction, this suggests rather that there is still a separation of mobility and play spaces, and that the primary function of mobility spaces is to enable efficient access to play spaces, rather than accommodate children's play as streets once did.

This rendering invisible of children's valuations of their mobility is exemplified in the focus on journeys to school in the literature (Carver *et al.* 2008), in particular the idea that the school journey is analogous to a 'commuting' journey that should therefore be efficient, productive and subject to conduct. One example of this is the development of 'Safer Routes to School' programmes in the UK that typically focus upon 'a combination of engineering, education and training, and other initiatives to reduce dependence on the car for the journey to school' (Bradshaw 1999, in Barker 2003: 137). The forms of mobility encouraged as part of these programmes – walking buses, car-sharing and cycle trains (Barker 2003: 137) – are evidently more about safety and efficiency than any notion of play or leisure. This focus on efficiency in the school journey is reinforced in a study by Carver *et al.* (2013). Here geographical information systems were used to record the most direct route along roads and footpaths from home to school for all children in the study (Carver *et al.* 2013: 465). The emphasis on directness assumes that children are 'rational' actors and take the shortest route possible. In making this assumption, such a methodology effectively frames out any playful or non-instrumental reasoning by assuming and superimposing an 'adult' rationality of efficiency onto children.

What qualities do children value when cycling?

One of the key things that comes across in this brief review of design guidance is that mobility spaces are not seen as places where children belong. Mobility spaces, it seems, should only facilitate adult/efficient movements and strictly adhere to and reproduce an understanding of the city street that seeks to accommodate life at a pace beyond the 'social average' (Harvey 1991: 229). This contrasts with the few accounts that have explored the desires and needs of children in relation to cycling. The pioneering urbanist Jan Gehl once noted that 'there is more to walking than walking' (quoted in Stafford and Baldwin 2018: 21). To paraphrase Gehl, what little literature that exists on children and cycling suggests that there is more to cycling – and particularly children's cycling – than just cycling.

Children's valuations of their own mobility prioritise its ludic and social elements. According to Oliver *et al.* (2016):

> children's travel behaviours tend to be less structured than more purposeful adult activity and travel patterns. When a child is walking with an adult they travel faster and more linearly and purposively compared with when they travel alone or with peers. Without journey supervision, children are free to explore, play and undertake challenging activities.
> (Oliver *et al.* 2016: 2)

This conclusion is reinforced in Horton *et al.*'s (2014) research considering the walking practices of 9–16-year-olds in the UK. Here they emphasise its spatial-temporal characteristics, noting its 'boundedness, intensity and circuitousness' (101). They also note its 'characteristic sociality, narrativity, playfulness and taken-for-grantedness' (101). What this points to again is that in contrast to adult valuations, the meanings of mobility for children are defined by a relative lack of 'function' in an instrumental sense, contributing rather to processes of identity formation, social bonding, making sense of the world and sensory pleasure. Indeed, as Horton *et al.* (2014) note, walking is not, 'for these children and young people, most importantly an instrumental means of getting "from A-to-B"' (104). Horton *et al.* go on to highlight (amongst other non-instrumental factors) the importance of play: 'many participants explicitly described their walking practices as a form of play. That is, they were often not setting out to play, or walking to play spaces, but walking itself was portrayed as enjoyable and playful per se' (109–110).

Such valuations are also present in research on children's cycling. Jay *et al.* (2009), in their exploration of the effects of public artwork installations on young people's valuations of using cycle routes, found that artworks participants found to be 'interactive' or 'fun' gave an enriched experience of using the route and helped to 'alleviate some of the boredom of cycling along a route' (Jay *et al.* 2009: 11, in Clayton and Musselwhite 2013: 2). Certainly, the fact that children might be 'bored' by cycle routes deemed suitable for adults speaks volumes about how differently children and adults approach cycling as a form of mobility.

Clayton and Musselwhite (2013) set out to explore what possibilities there might be to enhance the 'intrinsic rewards' (defined as excitement and fun) of cycling amongst young people and families. Whilst the methodology employed in their study often meant adults speaking for children (with all the caveats that implies), the study emphasised the benefits of designing more playful and interesting cycle routes. Clayton and Musselwhite comment that 'such additions might be a motivator for children to use cycle networks more frequently and independently, to use it for fun as opposed to simply for access' (2013: 12). Clayton and Musselwhite go on to suggest that such play and child-centred thinking could not only 'tap into children's desires for, and exciting sensory experience of using their bikes', but also 'make cycle paths *somewhere to go*, as opposed to simply something to use' (Clayton and Musselwhite 2013: 13). In contrast to current guidance, these studies foreground the fact that children are not 'incompetent adults', but rather that they have different goals and value different qualities than those prioritised (by the adult designers) in existing design guidance.

We can also problematise the reduction of children's mobility to just walking and cycling in such accounts. Lorimer and Marshall (2015) discuss this reductive tendency in the context of a broader idea of 'human locomotion', which they argue should include consideration of 'travel by small-wheel modes: inline skates, skateboards and push scooters' (1) and which have the potential to help create more 'agile' cities. As they go on to note, 'young people predominate in the use of alternative and non-traditional modes of transport' (such as small-wheeled modes) and that, as a result, these modes represent a 'new and innovative way of encouraging active travel among the young' (2). Lorimer and Marshall also suggest that such modes have potential beyond play. Somewhat surprisingly, Lorimer and Marshall's pilot study on the potential of using small-wheel modes as urban transport found that these were faster than public transport and walking for many journeys (2015: 5), and yet urban design has tended to actively exclude these modes – often favoured by children – rather than seek to accommodate them as serious forms of movement (Borden 2001; Woolley and Johns 2001; Woolley *et al.* 2011; Vivoni 2009).

The picture that emerges from such accounts is that children's cycling (and moving more generally) as a practice of place-making, social formation and identity construction/negotiation sits at odds with guidance that views it as 'just movement' and accordingly seeks to enhance only aspects such as safety and directness within rigidly defined (adult) parameters. Evidently in design guidance for children's active mobility, it is instrumentalised and adulterated, reducing it to a narrowly conceived idea of cycling that can contribute to positive health outcomes and maximised flows rather than as something to be encouraged in its own right. This position differs markedly with the views of children themselves, who prioritise the affective, playful and social elements of mobility over instrumental elements of convenience and efficiency. In a study commissioned by Ipsos-Mori, for example, when parents were asked what they deemed most important regarding cycling, 66% stated enjoyment. However, 77% thought their children would deem this most important (2010: 65). When children were asked this question, 89% stated enjoyment was most important (2010: 79). Conversely, 70% of adults associated cycling with health, whilst only 38% of children did so (2010: 65, 79). The disconnects here point to a partial instrumentalisation of cycling in adult life as a vehicle to achieve external goals, rather than children's readings of it as (a playful) end in itself. This chimes with wider literatures on play that draw attention to its marginalisation in urban life

(Spencer 2013; Stevens 2007), and research by Handy and Lee (2020 forthcoming) that foregrounds the importance of factors such as freedom and excitement in cycling, particularly for children.

The absence of children and indeed play in everyday mobility spaces is worth reflecting on because it actively produces children as lacking the required competencies at the same time as it produces these spaces as instrumental and adult. By way of example, if mobility spaces were designed with the goal of materialising and affording playfulness as a primary quality, this constructs the mobile subject as undirected, unproductive and 'childish'. If movement spaces are defined and materialised to prioritise the instrumental, they are likely to bring into being a mobile subject that is rational, productive and 'adult', and indeed actively exclude practices and subjects that do not conform. Thus following Barker *et al.*, we see 'age (young age) and mobility recursively produce one another in the course of young people's everyday lives' (Barker *et al.* 2009: 5). Accordingly the scripting of discourses around age in mobility spaces 'becomes internalized and lays latent in our behaviour towards people in the respective categories' (Goffman 1957: 14, in Frisk *et al.* 2016: 12). Such 'typifications translate into some pre-arranged roles and expectations for how to act, as old or young, with the behaviour that belongs to that category, which will determine how we act in the situation' (Frisk *et al.* 2016: 13).

A prime example of this can be found in mobility spaces, and particularly cycling infrastructure, where the understanding that adult cyclists are not playful in any way is reinforced in space, excluding more recreational activities by adults and young people alike. Age then is not 'simply, a "structuring" social category that controls children's (im)mobilities (nor is the reverse true). Rather, age and mobility are constantly produced in and through the experiential plane(s) of everyday life' (Barker *et al.* 2009: 5) mediated by design guidance and the infrastructures that flow from it. The point I want to emphasise here is that assumptions at the design stage around what cycling could be and how it is performed become materialised in environments that exclude certain user groups (such as children). Such exclusion is not only physical, as children could still use many of these spaces. The exclusion is rather also ontological because children do not see their priorities reflected in these designs and therefore may 'choose' to exclude themselves by not participating. It is little surprise that the subject of the child-cyclist has failed to emerge in the UK (and indeed most countries) given an inability to engage with and materialise their desires and affordances in mobility infrastructures.

Constructing the non-standard user in cycle design guidance

If children have been marginalised in cycle design, spare a thought for those with impairments. If children are effectively excluded because commuter infrastructures – whether designed for cars or bikes – prioritise speed and therefore physically and ontologically exclude them, those with impairments find that almost all aspects of current design exclude them because they are likely to require a non-standard cycle (such as a tricycle, hand bike or recumbent bicycle). For example, in a discussion of his use of non-standard cycles, Williams (2016) – suffering from development co-ordination disorder, affecting his balance and motor skills – discussed his transition from a standard bicycle to a recumbent tricycle. He notes that he favoured the non-standard tricycle on the basis that it 'took away all that need of balance' by giving a lower centre of gravity and extra wheel (NHS 2016, in Williams 2016). As this suggests, those with motor, balance and leg impairments are much more likely to require an adapted 'non-standard' cycle in order to participate in cycling as a practice.

A brief example on cycle lane widths from the *Welsh Active Travel Design Guidance* (ATDG) (2014) exemplifies a systemic lack of consideration and inconsistency when referring to non-standard cycles. ATDG – alongside the very similar *London Cycle Design Guidance* (2014) – acknowledges the need to consider non-standard cycles. Rather than consider each individually, ATDG suggests design parameters should aim to accommodate a 'standard inclusive cycle' with a width of 1.4 m and length of 2.7 m as representative of all non-standard cycle variations – trikes, cargo cycles, hand cycles, etc. (2014: 48). Based upon this, it recommends that cycle lanes are at least 1.5 m wide compared with the recommended 1.4 m for a standard cycle so that infrastructure does not exclude users of a non-standard cycle. However, the rationale for the additional 10 cm lane width is opaque at best given the potential difference in width between the standard and non-standard cycle. According to the guidance, assuming a width of 0.8 m for a standard upright cycle, a dynamic envelope (width) of 1 m should be allowed (2014: 47), yielding a ratio of 1.25 for the appropriate envelope. If this ratio is applied to the standard inclusive cycle width of 1.4 m, the minimum cycle lane width would be 1.75 m, not 1.5 m as stated. Moreover, this envelope assumes uniformly flat terrain (i.e. not the fact that a rider might wobble more on a hill start and therefore require a greater envelope) and that users of non-standard cycles would have the same level of control at slow speeds.

The significance is that even in what is seen to be good practice design guidance, the minimum specified width falls well short of what would be required to accommodate the standard inclusive cycle even based upon its own (opaque) criteria for calculation in ideal conditions. If we then assume a slower speed, a gradient, increased traffic, different turning circles, and the desire to facilitate social cycling (ATDG 2014: 55), the minimum lane width required to safely and comfortably facilitate these would most likely be greater still. Indeed, the minimum cycle lane width recommended for safe overtaking and social cycling of 2.5 m based upon a 'standard' dynamic envelope of 1 m (2014: 55) yields a ratio of 2.5 and therefore a minimum lane width of 3.5 m if applied to the 1.4 m standard inclusive cycle. This becomes all the more alarming when considering that many cycle lanes

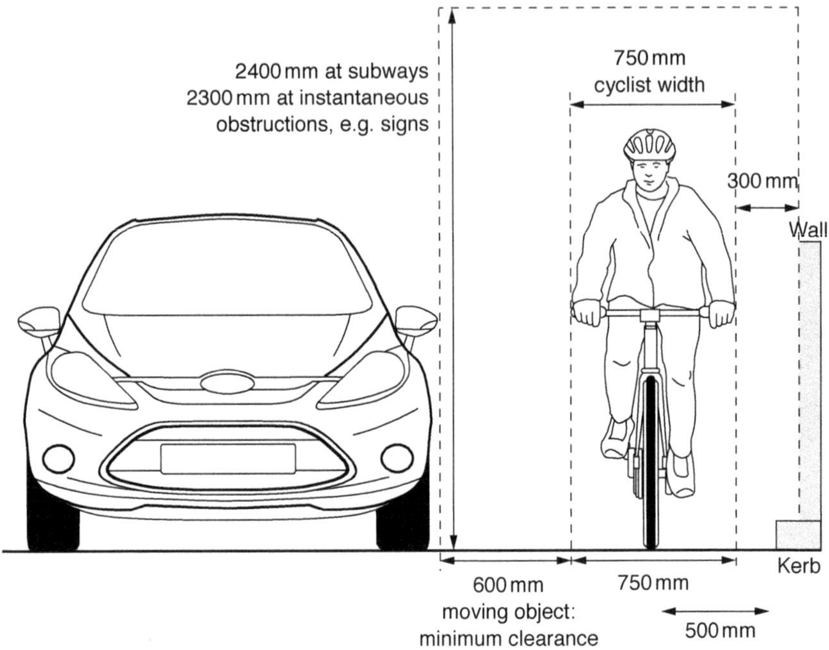

Figure 3.1 Basic space requirements for cycling.
Source: ATDG (2014: 54).

(old and new) do not meet even the basic 1.5 m width, let alone the 2.5 m stipulated for social cycling, or the potential 3.5 m minimum estimated here for the standard 1.4 m inclusive cycle. As Clayton *et al.* (2017) conclude:

> adapted cycles have dimensions which are often larger than those of a typical bicycle design, and barriers can be created when the design guidance which is used by engineers in the construction of infrastructure does not sufficiently take these dimensions into account.
>
> (454)

If this discussion of widths may seem esoteric and myopic (and is arguably both), it serves to exemplify the ways in which the affordances of non-standard cycle users are discounted in current design guidance through a series of assumptions. There are a number of other inconsistencies and normative assumptions in the ATDG (and LCDS) that could be interrogated further, including how it deals with issues of visibility, speed, gradients, barriers and turning circles (ATDG 2014: 47–53) in relation to non-standard cycles, alongside issues that are completely absent, such as the ways in which non-standard cycle users experience road cambers.

A study of disability cycling by Williams (2016) investigated many of these issues in relation to non-standard cycles. Williams found that surface quality became a key issue for his participants:

> for upright tri-cyclists, with a higher centre of gravity, and for riders with 'limited core muscles', poor surfacing induced greater precarity and danger from 'being thrown off balance', whereas recumbent tri-cyclists, who rode closer to the surface, were mindful of receiving injury, such as 'whiplash' from sudden bodily movements. The faster that cycle-riders negotiated poor quality surfaces, the more pronounced negative sensations appeared to be.
>
> (Williams 2016)

Similarly, heavy cambers designed with cars in mind, and easy enough to negotiate on a two-wheeled bicycle, were reported as exceptionally unwelcome. Whether riding a recumbent or tricycle, participants noted that cambers made their non-standard bikes steer into the kerb, which was accompanied by a loss of balance and a struggle to control the bike. This was articulated as unpleasant, unsafe and also meant that riders could not look around them to take in any potentially nice views because they were focused on staying safe:

> it means that you're having to be aware of the fact you're cycling, you can't just enjoy the activity, you have to think about every motion and every movement and every bit of steering. The bike wants to smash you into the kerb because of the camber.
>
> (Ford, recumbent tri-cyclist, quoted in Williams 2016)

As this suggests, even if they may remain negotiable, unsupportive cycling environments diminish the cycling experience.

Whilst it might be assumed that separated cycle lanes would offer non-standard cyclists more accommodating spaces, these were also heavily critiqued for 'entrapment' (Williams 2016). This was because of the fact that many non-standard users were unable to lift their bikes or perform hopping manoeuvres that were implicitly required when negotiating the physical boundaries that tended to separate cycle lanes from pedestrian and roads infrastructure (e.g. high kerbs). As a result, participants reported feeling 'trapped' in such infrastructure should they need to move out of it for any reason (Williams 2016).

Clayton *et al.* (2017) echo these findings, suggesting that whilst there are great potential health and wellbeing benefits from including those with impairments in cycling, in addition to the substantial cost of purchasing a non-standard cycle (up to £6,000), physical barriers remain because differently impaired users are not accounted for in design guidance or the resulting physical cycle infrastructure (461). Such absence limits accessibility for those who may wish to use non-standard cycles by both prohibiting participation (accessibility as outcome) and reducing the ease and enjoyment of mobility (accessibility as process) (Middleton and Spinney 2019).

Oudshoorn *et al.* (2004, 2016) have argued persuasively that one of the enduring reasons for such exclusion in design communities is because they remain dominated by white, able-bodied men in their thirties and forties whose views become embedded in methodologies and objects. Oudshoorn *et al.* (2016) argue that one of the real dangers of this absence of diversity in design professions is that even when other users are considered, they tend to be made up from stereotypes of what social groups are like (Oudshoorn *et al.* 2016: 171). Xie and Spinney (2018), for example, have demonstrated how gender-blindness from lack of diversity in transport planning professions results in gender-blind methodologies and ultimately infrastructure. The same process would appear to be at work in relation to children and the disabled, where it is assumed that these groups of users want the same as other users, or perhaps worse, are constituted as a problem group (e.g. dependent or incompetent other) rather than a group whose needs are legitimate and should be taken seriously (Oudshoorn *et al.* 2016: 172).

Conclusions

In this chapter I have demonstrated how the design practices and tools of cycle planners and engineers help bring into being a particular (adult, male, able-bodied, commuting) cycling subject whilst making it less likely (though not out of the question) that other cycling subjects and performances are likely to emerge. I have conceptualised design and its material outcomes as forms of ordering in that they are based upon ideas of how the world 'ought to be' and attempt to manage and control phenomena to stabilise and materialise these understandings. As Weicker and Sgibnev (2019) have noted, 'although mostly perceived as something stable in everyday life, at second glance, infrastructures consist of multiple layers with varying durability, changing over time and space on a material as well as a conceptual level' (196).

I have shown here that the tools, guidance and resulting infrastructure fail to take into account alternative uses of space and differentiated user affordances with the result that design guidance and resulting spaces are premised – where they consider cyclists at all – upon the stereotype of a physically capable, unencumbered and instrumentalised user in mind. It would appear that many users are not able to either physically or ontologically place themselves in this frame and thus are excluded from participation. As Cupples and Ridley (2008) note, such exclusion is likely to continue unless cycling policy and practice acknowledges 'the difficulties everyday transport might pose for those with different locations, bodies, social responsibilities or levels of income' (Cupples and Ridley 2008: 263). Hohne (2015: 314) argues that it is easy to forget that urban infrastructures are powerful devices for establishing and maintaining subject formations. As outlined in Chapter 2, the production and normalisation of a desired cycling subject and its integration into social, economic and cultural orders relies not only on its interpellation through social institutions and norms, it also relies on materialities.

If, as Miller (2002) argues, successful design is premised upon knowing the user, it would appear that cycle design needs to get to know a much wider range of users. However, the fact that it currently doesn't reinforces the thesis that contemporary cycle promotion is more concerned with accommodating economically 'productive' journeys rather than those that might be undertaken to facilitate play, identity-creation and freedom, despite the centrality of these to what makes cycling meaningful (Handy and Lee 2020). Writing on sustainable communities, Kraftl (2014: 275) acknowledges that whilst architecture, urban design and planning have long been concerned to govern life, they have featured little in discussions of bio-power. Indeed, Kraftl suggests these disciplines should be a foci of study in bio-politics precisely because the projects that emanate from them seek to articulate and intensify a focus on life itself and bring into being specific kinds of self-governing subjects (Kraftl 2014: 275). As Kraftl (2015) notes, through attempting to bring this subject into being, evidently certain kinds of life have been rendered visible, whilst others have been rendered invisible (220). Very clearly, playful, alternative and non-productive forms of life are increasingly designed out of public, and especially movement, spaces, in favour of more 'rational' and productive forms of life and comportment. Here we have to see the emphasis on active commuting as a form of 'anatamo-politics' that imports a number of assumptions and attendant exclusions: that walking and cycling are the most appropriate modes of active commuting; that commuting should be instrumental, not playful; that active commuting is an (able-bodied) adult practice; that the bikes to be ridden should be standard upright cycles. It is difficult to escape the conclusion that those deemed unproductive have been designed out of not only public transport systems, but now cycling systems as well.

Note

1 This view has some merit but also neglects to acknowledge the fact that countries like Denmark and the Netherlands have much higher gender equity generally, with supportive childcare regimes, meaning that women are not as encumbered by childcare duties as their UK counterparts. Hence, more accommodating design of infrastructure and bikes is only part of the solution.

References

Akrich, M. (1992) 'The description of technical objects', in Bijker, W. and Law, J. (eds) *Shaping Technology/Building Society: Studies in sociotechnical change* (Cambridge, MA: MIT Press), pp. 205–224.

Aldred, R. *et al.* (2016) Diversifying and normalising cycling in London, UK: An exploratory study on the influence of infrastructure, *Journal of Transport and Health*, 4, pp. 348–362.

Bannister, D. (2002) *Transport Planning* (London: Spon Press).

Barker, J. (2003) Passengers or political actors? Children's participation in transport policy and the micro political geographies of the family, *Space and Polity*, 7(2), pp. 135–151.

Barker, J. (2008) Methodologies for change? A critique of applied research in children's geographies, *Children's Geographies*, 6(2), pp. 183–194.

Barker, J., Kraftl, P., Horton, J. And Tucker, F. (2009) The road less travelled: New directions in children's and young people's mobility, *Mobilities*, 4(1), pp. 1–10,

Bendixson, T. (1974) *Instead of Cars* (London: Temple Smith).

Borden, I. (2001) *Skateboarding, Space and the City: Architecture and the body* (Oxford: Berg).

Buchanan, M., Bursey, N., Lewis, K. and Mullen, P. (1980) *Transport Planning for Greater London* (Hampshire, UK: Saxon House).

Carver, A., Timperio, A. and Crawford, D. (2008) Neighbourhood road environments and physical activity among youth: The CLAN Study, *Journal of Urban Health*, 85(4), pp. 532–544.

Carver, A., Watson, B., Shaw, B. and Hillman, M. (2013) A comparison study of children's independent mobility in England and Australia, *Children's Geographies*, 11(4), pp. 461–475.

Clayton, W. and Musselwhite, C. (2013) Exploring changes to cycle infrastructure to improve the experience of cycling for families, *Journal of Transport Geography*, 33, pp. 54–61. doi:10.1016/j.jtrangeo.2013.09.003.

Clayton, W., Parkin, J. and Billington, C. (2017) Cycling and disability: A call for further research, *Journal of Transport and Health*, 6, pp. 452–462.

Cresswell, T. (2006) *On the Move* (Oxford: Routledge).

Cupples, J. and Ridley, E. (2008) Towards a heterogeneous environmental responsibility: Sustainability and cycling fundamentalism, *AREA*, 40(2), pp. 254–264.

Dant, T. (2004) The driver-car, *Theory, Culture and Society*, 21(4–5), pp. 61–79.

Department for Transport (2017) *Walking and Cycling Statistics, England 2017*. Available at: https://assets.publishing.service.gov.uk/government/uploads/system/uploads/attachment_data/file/821842/walking-and-cycling-statistics-2018-accessible.pdf [accessed November 2019].

Department for Work and Pensions (2014) *Disability Facts and Figures*. Available at: www.gov.uk/government/publications/disability-facts-and-figures/disability-facts-and-figures [accessed August 2019].

Design Manual for Bicycle Traffic (2007) (Netherlands: CROW).

Emond, C.R., Tang, W. and Handy, S.L. (2009) *Explaining Gender Difference in Bicycling Behavior*, 88th Annual Meeting of the Transportation Research Board (Washington, DC: Transportation Research Board).

Fallan, K. (2008) Describing design: Appropriating script analysis to design history, *Design Issues*, 24(4), pp. 61–75.

Forstorp, P. (2006) Quantifying automobility: Speed, 'zero tolerance' and democracy, *Sociological Review*, 54(S1), pp. 93–112.

Franklin, A. (2004) Tourism as an ordering, *Tourist Studies*, 4(3), pp. 277–301.

Frisk, R., Frisk, J. and Due, T. (2016) *Designing Age-Integrated Cities: A framework for design exemplified with tweens*, paper presented at the 8th Making Cities Liveable Conference, Melbourne, 27–28 June.
Fullerton, B. (1975) *The Development of British Transport Networks* (Oxford: Oxford University Press).
Fyfe, N. (1998) 'Reading the street', in Fyfe, N. (ed.) *Images of the Street: Planning, identity and control in public space* (London: Routledge), pp. 1–10.
Garrard, J., Crawford, S. and Hakman, N. (2006) *Revolutions for Women: Increasing women's participation in cycling for recreation and transport: Summary of key findings* (Victoria, Australia: School of Health and Social Development, Deakin University).
Garrard, J., Handy, S. and Dill, J. (2012) 'Women and cycling', in Pucher, J. and Buehler, R. (eds) *City Cycling* (Cambridge, MA: MIT Press), pp. 211–234.
Gehl, J. (2001) *Life Between Buildings* (Copenhagen: Danish Architectural Press).
Handy, S. and Lee, A. (2020) What is it about bicycling? Evidence from Davis, California, *Travel Behaviour and Society*, 20, pp. 348–357.
Harvey, D. (1991) *The Condition of Post-Modernity: An enquiry into the origins of cultural change* (Malden, MA: Wiley- Blackwell).
Höhne, S. (2015) The birth of the urban passenger: Infrastructural subjectivity and the opening of the New York City subway, *City*, 19(2–3), pp. 313–321, doi: 10.1080/13604813.2015.1015276.
Horton, J., Christensen, P., Kraftl, P. and Hadfield-Hill, S. (2014) 'Walking ... just walking': how children and young people's everyday pedestrian practices matter, *Social and Cultural Geography*, 15(1), pp. 94–115.
Hunt, A. and Wickham, G. (1994) *Foucault and Law: Towards a sociology of law as governance* (London: Pluto Press).
Imrie, R. (2000) Disabling environments and the geography of access policies and practices, *Disability and Society*, 15(1), pp. 5–24.
Imrie, R. (2001) Barriered and bounded places and the spatialities of disability, *Urban Studies*, 38(2), pp. 231–237.
Ingold, T. (2000) *The Perception of the Environment: Essays in livelihood, dwelling and enskillment* (London: Routledge).
Ipsos-Mori (2010) *Research to Explore Perceptions and Experiences of Bikeability Training amongst Parents and Children* (London: Ipsos-Mori).
Jay, T., Mahdjoubi, L., Greene, L. and Walton, K. (2009) *A Toolkit for the Evaluation of Routes and Public Artworks with Young People* (unpublished).
Jones, T., Chatterjee, K., Spinney, J., Street, E., Van Reekum, C., Spencer, B., Jones, H., Leyland, L.A., Mann, C., Williams, S. and Beale, N. (2016) *Cycle BOOM: Design for lifelong health and wellbeing: Summary of key findings and recommendations* (Oxford: Oxford Brookes University).
Koglin, T. (2011) Planning for cycling = planning for equity: A response to Cupples and Ridley 'Towards a heterogeneous environmental responsibility: sustainability and cycling fundamentalism' (2008), *Area*, 43(2), pp. 225–227.
Kraftl, P. (2014) Liveability and urban architectures: Mol(ecul)ar biopower and the 'becoming lively' of sustainable communities, *Environment and Planning D: Society and Space*, 32, pp. 274–292.
Kraftl, P. (2015) Alter-childhoods: Biopolitics and childhoods in alternative education spaces, *Annals of the Association of American Geographers*, 105(1), pp. 219–237.
Krizek, K.J., Johnson, P.J. and Tilahun, N. (2005) *Gender Differences in Bicycling Behaviour and Facility Preferences: Research on women's issues in transportation*,

Conference Proceedings 35, Technical Papers, vol. 2 (Washington, DC: Transportation Research Board).

Law, J. (1992) Notes on the theory of the actor-network: Ordering, strategy and heterogeneity, *Systems Practice*, 5(4), pp. 379–393.

Law, R. (1999) Beyond 'women and transport': Towards new geographies of gender and daily mobility, *Progress in Human Geography*, 23(4), pp. 567–588.

Lawson, B. (2006) *How Designers Think: The design process demystified* (London: Routledge).

Lee, B. and Menzies, G. (2016) *Playful Cities: A cycle friendly city is a child friendly city*. Available at: www.thecityofplay.co.uk/post/2016/10/30/play-full-playful-cities [accessed August 2018].

Local Transport Note 2/04: *Adjacent and Shared Use Facilities for Pedestrians and Cyclists*. Available at: www.dft.gov.uk/consultations/archive/2004/ltnwc/ltn204adjacentandsharedusefa1692 [accessed July 2005].

Lorimer, S. and Marshall, S. (2015) Beyond walking and cycling: Scoping small-wheel modes, *Proceedings of the Institution of Civil Engineers – Engineering Sustainability*, 169(2), pp. 58–66.

Merriman, P. (2006) 'Mirror, signal, manoeuvre': Assembling and governing the motorway driver in late 1950s Britain, *Sociological Review*, 54(s1), pp. 75–92.

Merriman, P. (2015) Mobilities I: Departures, *Progress in Human Geography*, 39(1), pp. 87–95.

Michaels, M. (2000a) These boots are made for walking…: Mundane technology, the body and human-environment relations, *Body and Society*, 6(3–4), pp. 107–126.

Michaels, M. (2000b) *Reconnecting Culture, Technology and Nature* (London: Routledge).

Middleton, J. and Spinney, J. (2019) 'Social inclusion, accessibility and emotional work', in Docherty, I. and Shaw, J. (eds) *Transport Matters* (Bristol: Policy Press).

Miller, D. (2002) Turning Callon the right way up, *Economy and Society*, 31(2), pp. 218–233.

Morton, H.V. (1933) *The Pageant of the Century* (London: Odhams Press).

Moudon, A.V. *et al.* (2005) Cycling and the built environment: A US perspective, *Transportation Research: Part D*, 10(3), pp. 245–261.

Murdoch, J. (1997) Inhuman/nonhuman/human: Actor-network theory and the potential for a non-dualistic and symmetrical perspective on nature and society, *Environment and Planning D: Society and Space*, 15(6), pp. 731–756.

Oliver, M., McPhee, J., Carroll, P., Ikeda, E., Mavoa, S., Mackay, L. *et al.* (2016) Neighbourhoods for active kids: Study protocol for a cross-sectional examination of neighbourhood features and children's physical activity, active travel, independent mobility and body size, *BMJ Open*, 6(8), p. e013377.

Oudshoorn, N., Rommes, E. and Stienstra, M. (2004) Configuring the user as everybody: Gender and design cultures in information and communication technologies, *Science, Technology and Human Values*, 29(1), pp. 30–63.

Oudshoorn, N., Neven, L. and Stienstra, M. (2016) How diversity gets lost: Age and gender in design practices of information and communication technologies, *Journal of Women and Aging*, 28(2), pp. 170–185.

Parkin, J. (2018) *Designing for Cycle Traffic: International principles and practice* (London: ICE Publishing).

Pucher, J. and Buehler, R. (2011) *Analysis of Bicycling Trends and Policies in Large North American Cities: Lessons for New York*, University Transportation and

Research Center, Region 2 Final Report. Available at: www.utrc2.org/publications/analysis-bicycling-trends-and-policies-large-north-american-cities-lessons-new-york [accessed 20 November 2014].

Pyer, M. and Tucker, F. (2017) 'With us, we, like, physically can't': Transport, mobility and the leisure experiences of teenage wheelchair users, *Mobilities*, 12(1), pp. 36–52, doi: 10.1080/17450101.2014.970390.

Schliwa, G., Armitage, R., Aziz, S., Evans, J. and Rhoades, J. (2015) Sustainable city logistics: Making cargo cycles viable for urban freight transport, *Research in Transportation Business and Management*, 15, pp. 50–57.

Sennett, R. (1994) *Flesh and Stone* (London: Faber & Faber).

Sheller, M. and Urry, J. (2006) The new mobilities paradigm, *Environment and Planning A*, 38, pp. 207–226.

Southworth, M. and Ben-Joseph, E. (2003) *Streets and the Shaping of Towns and Cities* (Washington, DC: Island Press).

Spencer, B. (2013) *Playful Public Places for Later Life: How can neighbourhood public open space provide opportunities for improving older people's quality of life by enabling play?*, unpublished PhD thesis (Faculty of Environment and Technology, University of the West of England, Bristol, UK).

Spinney, J. (2006) A place of sense: A kinaesthetic ethnography of cyclists on Mt Ventoux, *Environment and Planning D: Society and Space*, 24(5), pp. 709–732.

Spinney, J., Reimer S. and Pinch, P. (eds) (2017) *Mobilising Design* (Oxford: Routledge).

Stafford, L. and Baldwin, C. (2018) Planning walkable neighborhoods: Are we overlooking diversity in abilities and ages?, *Journal of Planning Literature*, 33(1), pp. 17–30.

Stevens, Q. (2004) Artificial waterfronts, *Urban Design International*, 14(1), pp. 3–21.

Stevens, Q. (2007) *The Ludic City: Exploring the potential of public spaces* (Oxford: Routledge).

Storni, C. (2012) Unpacking design practices: The notion of thing in the making of artifacts, *Science, Technology and Human Values*, 37(1), pp. 88–123.

Sustrans (2009) *Routes to Play: A guide for local authorities* (Bristol, UK: Sustrans).

Transport for London (2010) *Cycling Revolution London* (London: Transport for London).

Urry, J. (2000) *Sociology Beyond Societies: Mobilities for the twenty-first century* (London/New York: Routledge).

Van der Duim, R., Carina, R. and Johanesson, G.T. (2013) Ordering, materiality, and multiplicity: Enacting actor– network theory in tourism, *Tourist Studies*, 13(1), pp. 3–20.

Vivoni, F. (2009) Spots of spatial desire: Skateparks, skateplazas and urban politics, *Journal of Sport and Social Issues*, 33, pp. 130–149.

Weicker, T. and Sgibnev, W. (2019) '12 Infrastructures as fluidities: How marshrutkas help us to overcome static conceptions of road-based mobility service provision', in Taukivene, T. Sgibnev, W. and Neugebauer, C. (eds) *Post-socialist Urban Infrastructures* (Oxford: Routledge), pp. 195–212.

Welsh Active Travel Deisgn Guidance (2014) *Design Guidance: Active Travel Wales Act 2013* (Cardiff: Welsh Government). Available at: https://gov.wales/active-travel-design-guidance [accessed November 2018].

Williams, S. (2016) *Thinking Beyond Two Wheels: Urban cycling persectives of adaptable cycle users*, unpublished MSc thesis (Cardiff University).

Winters, M., Friesen, M.C., Koehoorn, M. and Teschke, K. (2007) Utilitarian bicycling: A multilevel analysis of climate and personal influences, *American Journal of Preventive Medicine*, 32(1), pp. 52–58.
Woolley, H., Hazelwood, T. and Simkins, I. (2011) Don't skate here: Exclusion of skateboarders from urban civic spaces in three northern cities in England, *Journal of Urban Design*, 16(4), pp. 471–487.
Woolley, H. and Johns, R. (2001) Skateboarding: The city as a playground, *Journal of Urban Design*, 4, pp. 255–282.
Xie, L. and Spinney, J. (2018) 'I won't cycle on a route like this; I don't think I fully understood what isolation meant': A critical evaluation of the safety principles in Cycling Level of Service (CLoS) tools from a gender perspective, *Travel Behaviour and Society*, 13, pp. 197–213.
Yaneva, A. (2009) Making the social hold: Towards an actor-network theory of design, *Design and Culture*, 1, pp. 273–288.
Yaneva, A. (2012) *Mapping Controversies in Architecture* (London: Routledge).

4 Extracting surplus value from mobility[1]

Cycling policy and practice in London (UK) as a mode of political-economic and bio-political governance

Introduction

Chapter 2 concluded that despite recent positive developments, both transport and mobilities approaches to cycling have tended to be vague 'when dealing with issues of power, hegemony and social order' (Baerenholdt 2013: 21). In relation to cycling, for example, we've seen many papers (my own included) which have looked at what cycling as a practice means to those who do it as a sensory engagement (Jones 2012; Jungnickel and Aldred 2014; Nixon 2012; Spinney 2006, 2007); as a subcultural practice (Fincham 2006; Furness 2005); a gendered practice (Cresswell and Uteng 2008); and as a response to health and environment concerns (Bohm *et al.* 2006; Bonham 2006; Button and Nijkamp 1997; Newman *et al.* 1995). Whilst insightful in excavating a micro-politics of practice, it is fair to say that these accounts have failed to critically engage with the relationship between cycling and the neoliberal state.

Whilst it is tempting to see cycling as 'apple pie', as Stehlin (2014) has argued, contemporary cycling advocacy and practice must be situated critically within broader processes of capitalist urbanism (22). Taking London (UK) as the focus, this chapter seeks to do just that by connecting recent developments in cycling policy to forms of bio-political and political-economic governance. Some recent scholarship has examined the ways in which cycling policies are experienced by sub-altern and under-represented groups, and linked these to issues of method, wealth, policy, and planning to highlight the uneven and variegated nature of promotion and experience (Golub *et al.* 2016; Hoffman and Lugo 2014; Jennings 2015; Lam 2018; Leet *et al.* 2016; Meador 2016; Nixon 2012; Stehlin 2014, 2019). This chapter seeks to extend and complement this nascent literature by situating the growth of cycling in London (UK) within broader political-economic processes. In particular it seeks to understand why we see an emphasis on facilitating commuter cycling with less attention and little action devoted to other, less productive forms of cycling.

The first section explores the links between mobility and productivity in London and the inability of successive administrations to provide a spatial fix to decrease the circulation time of capital and labour (Harvey 1982). Drawing upon theoretical insights from Chapter 2, the section brings together insights on governmentality and bio-politics with political-economic theorisations of fixing to argue for seeing cycle promotion as a principally bio-political project that seeks to operate through shaping individuals as entrepreneurs of the self. In doing so, it argues that cycling policy can be viewed as a 'mobility fix' (Minn 2013) because current incarnations of cycling represent a way of governing and 'fixing' that works primarily through producing alternative ways of *moving* rather than space (Baerenholdt 2013). As this implies, mobility fixing is closely linked to 'nudge' and behaviour change strategies where 'states must govern from a distance, meaning that the only viable route for delivering certain policy goals, such as reductions in carbon emissions from personal mobility, is to encourage individuals to change their behavior' (Barr 2018: 3). I also argue that what little spatial fixing is evident to accompany such 'nudges' in the form of cycle-specific infrastructure (such as separated cycle lanes) can be understood as an improvised and contingent metaphorical fix (Jessop 2004) resulting from tensions created between attempts to mobilise the enterprise of cycling on the part of individuals and the realities of practising it in hostile urban environments.

In the second part of the chapter I situate a recent uptick of infrastructure roll-out (Lam 2018) as a kind of 'extraction architecture' (Zuboff 2019) serving to enlarge the scale and scope of cycling in relation to health, environmental and image issues as matters of concern. This places cycling in London firmly within debates regarding urban entrepreneurialism, arguing that beyond the energy required to get to work, the commuter cyclist also produces a 'behavioural surplus' because their labour power exceeds the labour time worked. That is to say, the value of cycling goes beyond its value to the individual or capitalist in getting a worker to work because of the ways it is associated with multiple matters of concern: the 'labour' of the cyclist therefore does not just enable space to be overcome; it helps to mitigate matters of ill-health because that labour is also rendered as exercise; environmental issues because that labour is low emission; and issues relating to place image because the labour of the cyclist can also be harnessed to project an image of liveability. As such, it is these 'latent' qualities of cycling (amonst many possible others) that are valorised because they are deemed to have a use value.

The chapter concludes by discussing some of the consequences of this neoliberalisation of urban cycling, particularly a narrow 'productivist' framing of cycling in the shape of commuter-focused infrastructure and promotion that effectively marginalises sub-altern and alternative performances (Hoffman and Lugo 2014; Stehlin 2014). The concluding hope of the chapter is that theorising the promotion and practice of cycling within broader processes of neoliberalisation (Tickell and Peck 2003) could help to direct future research agendas for cycling (and critical mobilities scholarship more broadly) that focus on issues of inclusivity, representation and inequality. In line with Cupples and Ridley (2008), the hope is that framing cycling in less productivist terms could encourage 'an aesthetics of a creative body in motion, where the rational geometric spaces of the city are turned into multiple lived spaces ... and where a fluid and affective ethical sensibility is working to undermine exclusionary and disciplinary forms of governmentalisation' (Cupples and Ridley 2008: 263).

Productivity, urban mobility and bio-politics

It is a well-rehearsed argument that since the 1930s, urban areas in Europe and North America began to experience de-industrialisation – processes that accelerated in the post-war period with increasing suburbanisation and the 'hollowing out' of cities (Hackworth 2007; Harvey 1991). As White (2008) points out in the case of London, alongside decentralisation policies for employment and industry, which emptied inner cities, came a substantial lack of investment and privatisation of public transport. In London this manifested, for example, in falling ridership on the London Underground, which reached a low of 498 million in 1982 (White 2008: 184–185). At the same time, the opening of the bus network to competitive tendering in 1986 began to produce a more uneven transport geography as providers focused on more profitable routes (Docherty and Shaw 2009). Evidently (dis)investment in transport infrastructures and systems mirrors that of broader spatial infrastructures.

From the 1970s onwards, inner-city areas of the UK, North America and Europe have been the focus of accelerating spatial fixes as administrations have offered incentives to revitalise inner cities, resulting in an influx of financial services and a patchy development of consumption and residential spaces in or near once faded central business districts (CBDs) (Hackworth 2007: 80) to remedy economic crises. As Lees (2003) has noted, 'over the last three decades schemes to attract the wealthy middle classes back to the inner city have become central to urban redevelopment strategies' (613).

As investment, businesses and people have been encouraged back into inner cities areas, neglected roads and mass transit systems have struggled to keep up with the increases in demand. As Figure 4.1 illustrates, morning commuter numbers in London peaked at about 1.25 million in 1962 and then declined fairly consistently over much of the next 20 years. Numbers began to pick up in the early 1980s, with peaks and troughs in the last 30 years, reflecting London's economic performance in a globalised economy. This suggests a clear link between processes of inner-city valorisation and the mobility systems that serve to circulate goods and labour.

In cities such as London, with historic under-investment in public transport (Docherty and Shaw 2009), this recuperation of inner areas has come at a cost. A combination of rising populations, increased demand for public transport, increased car ownership and a rise in commuting – in from surrounding

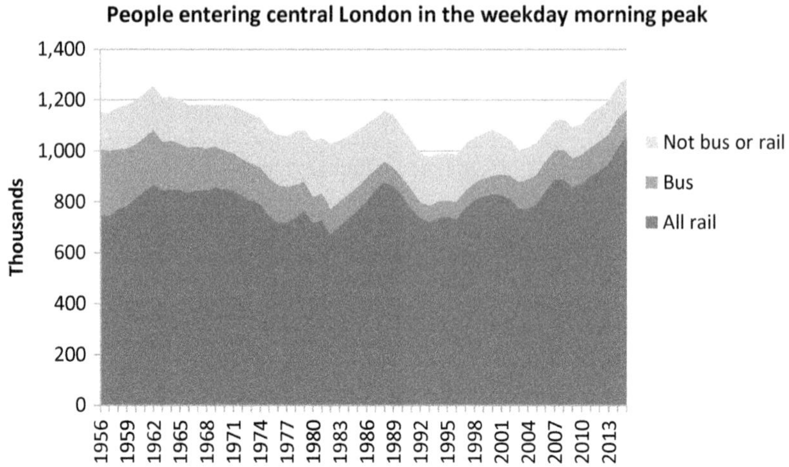

Figure 4.1 People entering central London in the weekday morning peak, 1978–2009.
Source: James Gleeson, http://londontransportdata.wordpress.com/2012/01/16/long-run-trend-in-commuting-into-central-london/.

counties has resulted in reduced transit times due to overcapacity. In London, as in many cities across the world, the 'success' of particular forms of mobility (notably the car system) has resulted in the circulation of capital and labour slowing down to the point where these systems of mobility reduce the rate of capital accumulation (Hickman and Bannister 2015).

Harvey's argument (rehearsed in Chapter 2) regarding the importance of mobility systems to capitalist economies is exemplified in cities like London, where, according to a recent report, 'companies located in Central London are highly dependent on the transport of staff' and to a lesser extent that of goods for their productivity (GLA Economics 2005: 53). However, average traffic speeds in the city have decreased consistently since the mid-1970s (GLA Economics 2005: 9) as businesses and residents have been attracted back to the centre and congestion has worsened. This trend of declining speeds has continued with recent TfL data from its core central London road network, showing average traffic speeds between 7am and 7pm of 7.4 mph in 2017/18.

A report by GLA Economics found that 'the overall minimum quantifiable cost of transport delays in Central London to commuters and businesses is £1,190 million per annum' (GLA Economics 2005: 32). Including costs for non-work purposes, the report concludes that the overall cost of congestion in London is £6.7 million per business day (ibid.), equivalent to around 1% of London's GDP (GLA Economics 2005: 52). For London, the revalorisation of inner-city areas, with its corresponding increase in the volume of goods and people to be moved within ageing and under-funded mobility spaces, has evidently come at the expense of efficiency in its mobility systems and therefore productivity.

A key problem highlighted by Harvey (1982) with regard to speeding up these now unproductive circulation times is that the physical infrastructures upon which mobility relies are 'immobile in space and highly vulnerable to place-specific devaluation' (Harvey 1982: 380). As Harvey points out, 'capitalism increasingly relies upon fixed capital (including that embedded in a specific landscape of production) to revolutionise the value productivity of labour, only to find that its fixity (the specific geographical distribution) becomes the barrier to be overcome' (Harvey 1982: 394). Indeed, Harvey goes on to argue that specifically it is the value embedded in the transport system – its 'sunk costs' – that becomes the main barrier 'to the further expansion of value in general' (Harvey 1982: 380).

While *et al.* (2004) suggest that cities are in a phase of eco-restructuring based on limiting carbon, which is giving rise to a 'distinctive political economy' that seeks to extend 'state intervention in the spheres of production and consumption' (While *et al.* 2004: 66). Bulkeley (2010) confirms that transportation is a key arena for such interventions:

> as the issue of reducing GHG emissions has started to be addressed in a more geographically diverse range of cities, there is some evidence of a switch from a concern with reducing the domestic use of energy to commercial buildings and of the development of initiatives that target transportation issues, primarily to address congestion and air quality.
>
> (Bulkeley 2010: 235)

Given the economic costs of increased pollution, a spatial fix in the shape of more road-building would – at least in the longer term – exacerbate problems of accumulation. A recent OECD report states that outdoor air pollution kills over three and a half million people worldwide every year. It goes on to state that in the UK, the economic cost of deaths from air pollution amounted to about £50 billion in 2010, equivalent to roughly half of government health expenditure (OECD 2014). Moreover, urban populations are disproportionately affected by such issues because of higher population densities in metropolitan areas (SQW Consulting 2008: 20). As scientific evidence and concern over local and global pollutants has grown, opposition to private motorised transport in cities has also mobilised. City administrations have been slow to take this on board, but the most recent TfL documents suggest it is attempting to shift away from private cars through promotion of bus and active travel (walking and cycling) as alternatives (TfL 2013).

A further notable environmental barrier to adding new road capacity relates to London as a heritage site, with many of London's buildings and frontages designated as UNESCO World Heritage sites (Doward 2013). As ongoing debates around tall buildings in central London indicate, the ability to construct either new buildings or roads around these sites is severely curtailed by the planning constraints that come with such designations. In addition, there is ongoing political fallout from the last round of major road 'trunking' in the 1960s, which effectively turned many roads into urban motorways (for example, the A406 North Circular and A13 in East London), severing and blighting local communities (Golbuff and Aldred 2011).

Accordingly successive London governments have grappled with the kinds of fixes required to decrease the circulation time of capital and labour in order to improve productivity, and encourage modal shift to address related matters of concern. In the present, for example, there is evidence of a spatial fix occurring in London in the form of Crossrail: the construction of a new East–West underground rail line. Whilst this is costly – £17.6 billion at the last reckoning (Briginshaw 2019) – its spatial impact on the surface is relatively limited: the vast majority of this spatial fix is subterranean. In contrast to the very visible UK road-trunking programme of the 1960s,[2] and 1980s policies geared towards road-building (e.g. Roads for Prosperity 1989), what has become abundantly apparent is that municipal administrations are increasingly unwilling to institute spatial fixes in the form of urban road-building.[3]

This reluctance is exemplified in the latest TfL *Roads Modernisation Plan*, which outlines £4 billion worth of road spending between 2014 and 2021 (2014: 3). The ambition of these spatial fixes is modest, however, with no new capacity planned; the £4 billion is to be spent on alterations to the existing network, including junction treatments, surface treatments, bridge treatments and signal upgrades. Apart from £1 billion to be spent on cycle infrastructure (discussed later), the £4 billion is little more than a maintenance budget for the existing network. Indeed, as the 2013 TfL RTF report states, 'London is approaching the limit of what can be achieved through road space reallocation alone' (2013: 8). In other words, London – like many other cities – has little appetite for terrestrial spatial fixes to the transport system. There are three principle reasons for this: first, and certainly against the backdrop of London's slow post-2008 economic crash recovery, neither the state nor private enterprise (if we can even still distinguish) are willing or able to take the 'massive devaluation' that wholesale rebuilding of the roads system would require in order to function significantly more efficiently;[4] second, such direct interventions are not favoured in neoliberal modes of governance, where approaches such as travel demand management and 'nudging' are preferred (Whitehead *et al.* 2017; Barr 2018); and third, because the notion that falling productivity is solely due to increasing circulation times of capital and labour is far from the whole story.

Human capital and urban cycle promotion as bio-political 'mobility' fix

It is evident that whilst sedentary forms of mobility, such as the car, bus[5] and train, have historically improved circulation times of labour (through decreased commute times), they have lowered its productivity in other ways by producing more sedentary and less healthy labourers (and producing pollutants that harm the health of workers). Indeed, as Callon (2007) points out, the function of markets (in this case a mobility market) can:

> trigger the emergence of matters of concern to which they are not always able to provide satisfactory answers. These matters of concern then evolve into many (potentially) political issues whose solutions may, in turn, impact on the organization of economic activities.
>
> (Callon 2007: 139)

By way of example, a recent report quantifying the economic costs of sedentary transport exemplifies the transformation of this matter of concern into new forms of economic activities, stating that:

> the average employee has 6.8 sick days per year in the UK. A conservative estimate is that increased exercise can lower this sick leave by 6%. The value of lowering sick leave by this 6% is equal to £47.68 per working age adult across England based on an annual GVA per employee of £37,000.
>
> (SQW Consulting 2008: 10)

In line with this monetisation of the impacts of sedentary travel, a key goal of UK transport policy in recent years has been to encourage more 'active' modes of travel, such as cycling and walking. Certainly a spatial fix to increase car use would not satisfy the health-economic argument for changes in travel behaviour.

As the health-economic argument suggests, productivity cannot be maximised through the speeding up of labour or capital alone. Foucault (2010) has argued that explanation must also turn to the ways in which human capital is augmented (or not) through mobility, and how this contributes to increases in productivity (232). As I have already outlined in Chapter 2, whilst often criticised for downplaying 'extra-economic' factors such as human capital, Harvey is not blind to this issue. Writing in 1989, Harvey suggests that one way in which new entrepreneurial forms of city governance manifest is through the creation and valorisation of new or existing forms of labour power (Harvey 1989: 9), what I have referred to in Chapter 2 as the 'reskilling' of the population. In the modern city this means, for example, not only bringing new skills to bear in the sphere of (paid) labour reproduction, but also encouraging new skills in the sphere of (unpaid) social reproduction, enabling (for example) a healthier workforce to emerge. The issue for municipal governments, then, is not just one of improving circulation time through fixing mobility, but also enhancing human capital through mobility in other ways. Harvey implies as much when he states that 'the problem of geographical mobility is converted into one of transformation of the social and physical environment within which other forms of capital circulate' (Harvey 1982: 406). As Cerny (2006: 690) goes on to point out, the competitive state[6] is not only one whose economic policies focus on promoting competitiveness, but one which attempts to incorporate competitiveness into everyday social reproduction. Accordingly Foucault argues that modern policies of growth are 'focused precisely on one of the things that the west can modify so easily, and that is the form of investment in human capital' (2010: 232). Accordingly, using Foucault, I argue here that we can see cycling as a form of self-government and 'entrepreneurship of the self' situated within a bio-politics of obesity and sedentarism.

Bio-politics is concerned with 'processes that sustain or retard the optimization of the life of a population' (Dean 2010: 119). These include birth rate, hygiene, health, etc. (Franklin 2008: 317, in Dean 2010: 118). It is important to draw attention to the term optimisation here, because it could mean many things, but in the context of liberal government, Dean argues that it refers to 'frugal government … of economic processes' (Dean 2010: 121). It is here that I want to introduce cycling in London – once championed only by 'mad radicals' such as the London Cycling Campaign and Friends of the Earth (Spinney 2008) – as exemplifying a 'frugal' mobility practice increasingly promoted as a way of enhancing urban productivity and competitiveness.

A 2008 report by SQW, *Planning for Cycling*, set out the economic argument to invest in cycling, stating that its overall value accrues from a unique combination of improvement in general health and fitness; reduced pollution and the emission of CO_2;[7] and help in tackling congestion. The study calculated that even a return to 1995 levels of cycling (still a very low level but an increase of 50% on existing figures) would provide health, pollution and congestion savings of more than £1.3 billion annually in England (SQW Consulting 2008: 2). This technical knowledge of a population is bio-political in that it is explicitly produced in order to act on and provide a rationality to optimise the population. Certainly, a recent report for the NHS – in line with other similar reports – argued that the health benefits of cycling are the largest single reason to promote it (Davis 2010: 8). In London, this position has been officially embodied in transport policy with the GLA's current stated goal to increase cycling levels to 5% mode share by 2025, which would mean an overall 400% increase (GLA 2014).[8]

In *The Birth of Bio-politics* (2010), Foucault argues that because neoliberalism pursues 'frugal government', it cannot intervene directly in the economic, and must intervene instead 'on society … in its fabric and depth' (2010: 145). In so doing it will become a government of enterprise society rather than economy, where the individual becomes the locus of 'enterprise and production' (2010: 146–147). Using the example of migration (though equally applicable to the kind of intra-urban mobility that cycling represents), Foucault goes on to discuss mobility directly in relation to enterprise, theorising mobility as an investment on the part of the individual because it leads to an increase in status and remuneration (2010: 230). Consequently, the individual becomes an investor; 'an entrepreneur of himself who incurs expenses by investing to obtain some kind of improvement' (ibid.). Thus for Foucault, the ways in which an individual moves are conceived as a form of enterprise because of the financial and bodily rewards of moving in particular ways. If we relate this to cycling, we can see that the extra energy and effort (and sometimes discomfort) required to commute by bike may be paid back to the cyclist in the form of feeling healthier, and perhaps saving money and time. The benefits that accrue are evidently both instrumental and affective.

However, it is also clear that the same active commute that provides a benefit to the individual (in getting from A to B and feeling 'healthier') also provides value to the state beyond any uplift in productivity from getting workers to work faster or more reliably. This capacity of the active commuter to produce an output that exceeds the work they have input into getting from A to B represents what Zuboff (2019) has termed 'behavioural surplus'. Situated within valorising economic imperatives to lower pollution, reduce circulation times and improve worker health, the ability to tap into previously untapped elements of the active commuter's labour power is generative of surplus value because of the additional value of this mobility work in mitigating multiple matters of concern. Far from only benefiting the individual, the extra work of cycling in this context is 'extracted ... as the means to others' ends' (Zuboff 2019: 10): in this case, to lower health costs through better air quality and more active citizens, and to boost place brand by projecting an image of liveability.

As Harvey has argued, in order to produce labour (or in this instance citizens more generally) with the requisite skills to be mobile in appropriate ways requires the disciplining of labour power in the form of labour control: 'a mix of repression, habituation, co-option and co-operation, all of which have to be organised ... throughout society at large' (1991: 123). Thus, for the individual to invest in themselves and gain the disposition and skills required to become a cycle commuter requires the construction of a social infrastructure serving to inculcate self-management and construct 'reflexive health entrepreneurs, willing and able to manage their own wellbeing under the guidance of "distant" experts' (Cramshaw 2012: 200). This is exemplified in strategy documents such as the GLA report *Cycling Revolution London* (2010). Despite reference to some 'hard' measures, cycle promotion in the report is mainly concerned with equipping individuals with the correct incentives and skills to cycle, including cycle training, workplace travel planning, wayfinding and general awareness and communication (2010: 1). Cycle promotion policies are primarily concerned with shaping social (as opposed to spatial) infrastructures to influence the individual to invest in themselves: 'as an enterprise to be formed in direct response to the imperatives of neo-liberal choice, itself provided through a system of markets and incentives to achieve self-actualisation through judicious navigation of a series of regulated freedoms' (Rose 1999).

TfL's cycling policy reflects a broader national trend, where, as Barr (2018) has demonstrated, the UK government has spent considerable sums of money to create 'national policy architectures', such as the Behavioural Insights Team, to support a 'shift toward individualized forms of behavior change as the invocation of the "citizen-consumer," an individual who simultaneously embodies diligent responsibilities whilst exercising these through choice, rather than through (regulated) necessity' (Barr 2018: 3). Rose (1999) has argued that neoliberalist government requires individuals to act upon themselves as both free and responsible in order to fully realise its goals (68) and limit its involvement. Numerous commentators have pursued Foucault's arguments since his 1978 lectures, and the tendency within Western democracies to individualise responsibility for wellbeing is well documented (Cramshaw 2012: 200; Greco 2009; Robertson 2001). Cramshaw (2012), for example, highlights the ways in which social marketing is increasingly used to problematise and target health behaviours and set behavioural goals with the aim of constructing 'entrepreneurial dispositions towards the construction of the healthy self' (201).

It is evident that situated within a wider constructed politics of obesity, sedentarism (Guthman 2009) and environmental degradation, cycling becomes a key medium through which the individual can invest in themselves and (hopefully) obtain an improvement in the form of vitality and increased life expectancy. At the same time, the state reaps the surplus value of this 'improved' and reskilled labour in the form of a reduction in sick days, decreased health care costs, reduced circulation time of labour, reduced environmental costs, and improved place brand image. As Cramshaw (2012) has noted in relation to recent public health strategies, these imperatives reflect a shift from the social to the individual as the target for programmes of self-discipline relating to diet and exercise, amongst others (202). Crucially, this is government acting not only on mobility through a set of desired outcomes, but government through mobility (Baerenholdt 2014): cycling becomes a terrain through which the individual recognises and becomes equipped to deal with obesity and sedentarism as public health problems. Hence as a type of 'fix', cycling is best theorised as a 'mobility fix' because it represents a way of governing that works primarily through the production of new modes of mobile comportment rather than the production of space.

Accordingly I want to suggest that we can conceptualise attempts to fix mobility through the production of cycling as a form of 'market making'. Evidently there is a process of singularisation (Callon *et al.* 2002) occurring, where, through the use of particular market devices (such as evidence of health, environmental and economic benefits), cycling is represented as possessing particular qualities that it is hoped will make it attractive to citizens, distinct from other modes in terms of its benefits, and a market they wish to 'buy' into. This process echoes that undertaken by marketing professionals in 'traditional' consumer markets, where manufacturers and retailers are active in 'defining, shaping, transforming, qualifying and re-qualifying products' (Entwistle 2006: 706). Here market professionals are active in attempting to shape the qualities of goods (for example in the case of bottled water) by positioning them within specific discourses, such as weather patterns, global instability and health (Azimont and Araujo 2007: 853–855). These processes of requalification result in particular configurations of products appearing on the 'shop floor' (Entwistle 2006) that attempt to redefine what is right for a given moment (e.g. fashionable) and thus prompt consumer judgements regarding the worth of what they already own. Evidently the requalification of cycling as healthy, fast and sustainable is a contested attempt (see, for example, Cupples and Ridley 2008) to define what is right for the current 'moment' in the hope that it will detach citizens from their existing unsustainable mobility patterns.

Extraction architectures: cycle superhighways and productivity

Whilst cycle promotion in London (and indeed many cities around the world) is clearly tied up with a broader bio-politics, Foucauldian analyses of governmental projects should not blind us to uncertainties in whether such projects manage to meet their ends. As Baerenholdt (2013) notes, discursive intent does not make the world (230), and the political landscape is littered with the failures of bio-political projects (Cramshaw 2012; Nettleton and Green 2014). On the face of it, the promotion of cycling in London would appear to have been successful so far, with a rapid increase over the past 15 years. Census data shows a 144% increase in cycling in London over the period 2001–2011 (ONS 2014), with a daily average of 131,000 cycle journeys in 2014 (GLA 2014). The most recent figures show that cycle journeys in June 2018 were 187,345 per day, an increase of 8% year on year (TfL 2018).

However, leaving aside those who do not respond to exhortations to cycle, even for those who do take up cycling, cycling is far from safe. In recent years, cyclist deaths (particularly under the wheels of heavy goods vehicles – HGVs[9]) and safety more generally have become an increasing cause for concern. A prominent theme in media reports of cyclist fatalities has been a preponderance of blame falling on the abstract systems of road and junction design. Pressured by an extremely bad two weeks in November 2013, when six cyclists died, TfL responded with a number of infrastructural initiatives. In addition to its ongoing commitment to a 'Central London Grid' of cycle infrastructure, 'combining 20 miles of arterial Superhighways on main roads with a broad 60-mile network of capillary Quietways on calmer backstreets' (TfL 2014: 52), in March 2014 TfL announced the new 'Mini-Holland programme'. According to TfL, this programme aims to emulate cycle-friendly, low-traffic neighbourhoods in Amsterdam.

The winning bids for this funding included 'redesigns of town centres, new suburban cycle superhighways, Dutch-style roundabouts' (TfL 2014: 59).

As I have already argued, until recently cycle promotion in London (and the UK more widely) has tended to be in the shape of 'soft' measures, as Pooley *et al.* (2013) confirm: 'recent policies have been reluctant to adopt more interventionist approaches relying instead on persuasion and promotion of active travel mainly on health grounds' (Pooley *et al.*, 2013: 67, see also Golbuff and Aldred 2011). Official consensus in the UK has until very recently been that cycling provision should not require separate infrastructural 'fixes' as expressed in the 'hierarchy of solutions' (CIHT 1996; DfT 2008). However, in its more recent phase, as Lam (2018) argues, London's cycle boom is 'characterised by unprecedented investment in cycling infrastructure' and promises of 'a proper network of cycle routes throughout the city' (GLA 2013: 4, in Lam 2018: 115). Lam goes on to argue that increasingly 'infrastructure foregrounds London's cycling policy discourse, particularly infrastructures of mobility that manifest as either Cycle Superhighways or Quietways'[10] (Lam 2018: 115).

This raft of measures is evidence that the loose alliance of TfL and its partners has increased its efforts to encourage cycling by producing space for it, mainly in the form of redesigning and reassigning existing transit space rather than the production of new capacity from other uses (e.g. from land zoned for parking, retail or residential). The rolling out of the superhighways network and 'Mini-Holland' programme must be seen as a direct response to tensions arising between the mobility fix of cycle promotion, and the inhospitable environments within which it is practised (cf. Spinney *et al.* 2015). Indeed, it is increasingly apparent that as awareness of the risks of cycling amongst vehicular traffic in London grows as more people do it (Spinney *et al.* 2015), these risks serve to 'push back' upon and produce new spatial configurations.

We can also see cycle infrastructure as an integral part of mobility fixing designed to bring into being certain kinds of cycling subjects and maximise behavioural surplus. According to Zuboff (2019), maximising the value of behavioural surplus requires economies of scale, scope and action. We can recognise these at work in the roll-out of cycle-specific infrastructure in London. First, scale: in order to realise the surplus value of cycling requires as many people to cycle as possible, as the Mayor of London's website clearly states: 'analysis by the Greater London Authority shows that if every Londoner walked or cycled for 20 minutes a day, it would save the NHS £1.7bn in treatment costs over the next 25 years' (London Assembly 2017). Such comments unequivocally point to the fact that quantity matters. In addition, it is clear that for London to present itself on the global stage as a green and liveable city requires citizens to perform this image in the form of cycling: the more the better. However, safety also matters (to both the individual and the state), so in order to encourage more people to cycle requires a safer and more convenient environment for them to do so. Hence, echoing McFall (2009), I argue here that cycle infrastructure should be conceptualised as a 'market device' because of the ways in which it formats 'the dispositions and skills of the people who encounter' it (2009: 277).

However, whilst the number of cyclists is important, behavioural surplus must also be varied (Zuboff 2019: 193–194), because it is important to get as wide a range of people cycling as possible so that economic benefits are not just being generated by a narrow portion of the population: the work of mobility should be spread across the population. However, as already noted in Chapters 2 and 3, London has not been particularly good at this, with women, children, elderly, disabled and ethnic minorities under-represented in the cycling renaissance (Aldred 2015). As Lam (2018) notes, even in high cycling areas of London like Hackney, with 15.4% of all commuter journeys undertaken by bike, only 37% of Hackney's cyclists are women (Lam 2018: 116). London's gender gap as a whole is even worse, with 74% of all cycling trips made by men (Aldred 2015: 46).

One of the ways in which London is trying to increase the scope of those who cycle, and in particular encourage more women to do it, is to build more separated and safer cycle lanes. Research suggests that women are more concerned about safety issues associated with cycling than their counterparts (Twaddle *et al.* 2010) and value cycle lanes more than men (Krizek *et al.* 2005). As a result, women's participation in cycling is viewed as an 'indicator' of a cycling-friendly culture and environment (Garrard *et al.* 2006: 11) and efforts to get more women cycling often focus on the creation of separated infrastructure. In particular, London's nascent quietways network can be seen as an attempt to not only enhance safety, but in doing so provide the means to cycle for non-commute social reproductive journeys at the neighbourhood level, which are more likely to be undertaken by women. It is notable, however, that not only is TfL's quietways programme much more modest in scope than that of its superhighways (TfL 2018, *FOI request*), at the time of writing only one quietway and sections of three others had been completed (TfL 2018, *Quietways*). Whilst cycle infrastructure (at least in theory) is also fundamentally concerned with achieving economies of scope in order to facilitate behavioural surplus, the fact that quietways have not been given priority in London speaks volumes about what its policy priorities are.

This brings us on to cycle infrastructure as constitutive of an economy of action, and in particular the kinds of action that are encouraged and discouraged. Huxley (2006) calls infrastructure a 'dispositional spatial rationality': according to Huxley, this kind of rationality aims at drawing boundaries and producing order that will foster correct comportments. It operates with the logics of grids of classification for the spatial disposition of 'men and things' to bring arrangement and visibility to bear on individuals and populations problematised as chaotic and uncontrolled (Huxley 2006: 774, in Cupples and Ridley 2008: 259). As this suggests, not only does cycle infrastructure attempt to enable and induce cycling as a behaviour, but it also attempts to direct and shape the qualities of cycling practice in certain directions to produce the right kinds of behaviours. Cycle superhighways in particular prioritise the performance of particular variants of cycling, most notably the fast, efficient and direct kind that can make the maximum impact on reducing circulation times and reducing expenditure on health and environmental mitigation, by being a viable alternative to other commute modes.

A focus on the routes and widths of cycle superhighways demonstrates how they seek to bring into being a particular kind of productive cycling subject rather than focus on other qualities of cycling. Cycle superhighways are primarily concerned to get workers into and out of London's CBD. The fact that roads around them are untreated for cycle safety improvements means that safe cycling can only be undertaken on these routes: deviation means that cyclists are once again 'swimming with the sharks', so to speak. It also means that the experience of cycling on many of the routes remains one of cycling in close proximity to large volumes of traffic and large vehicles.

The width of the cycle lanes is also telling regarding what they are attempting to facilitate. The lanes are generally at the minimum end of widths specified in the LCDS (around 1.5 m), particularly given the increasing volumes of cyclists on them. There are two key implications here regarding the range of cycling that can be practised. First, with primarily fast commuters using the routes, there is little scope for both slow and fast cyclists to be accommodated, with minimal room for overtaking. Such a situation is likely to disadvantage inexperienced, slow and younger cyclists, who may be put off using the routes even if they are conveniently located. The narrowness of the routes also means that more social variants of cycling are unlikely to materialise because cyclists cannot comfortably ride side by side. Whilst driving two abreast is officially mandated in the design of cars and carriageway, it seems that cycling two abreast remains something that should not be encouraged. Hence superhighways bring into being a particular version of cycling and corresponding cycling subjects that emphasises a narrow range of possible qualities and marginalises others (and the subjects who value them).

Productivity, sustainability and the narrowing of cycling

This narrowing of cycling is not something only seen in London. It is also evident in cities with high rates of cycling, as confirmed by Carstensen *et al.* (2015) in a discussion of the rationale underpinning Copenhagen's extensive network of separated cycle lanes, which include:

> fiscal incentives to cycle to work, development of appropriate carriage and integration arrangements with public transport, good bike-to-public transport transfer options … high-quality cycle parking, proper

> path maintenance and properly designed crossings … widening existing paths, implementing 'green waves for cyclists', and testing alternative intersection designs.
>
> (Carstensen *et al.* 2015: 143–153)

Echoing this narrative, when discussing Copenhagen's cycle culture and infrastructure, Nielsen *et al.* (2013) state that 'the bulk of cycling in Denmark is what has been called "utility" cycling and connects to travel destinations (>90% of total cycling distance) rather than being leisure oriented' (111). This is also reflected in the city's annual Bicycle Account, which only measures cycle flows in relation to work and education journeys.

Accordingly cycle superhighways must be conceptualised as spaces configured to 'nudge, tune, herd, manipulate, and modify behaviour in specific directions' (Zuboff 2019: 193–194). Whilst the outcome of this may benefit the individual, it also represents a narrowing of a potential plurality of cycling practices as aspects of it that are extraneous to or may detract from surplus value being extracted (such as play, protest, exploration) are written out of these cycling scripts in favour of those better able to address matters of concern. Cycle infrastructure can therefore be viewed as part of a 'pre-emptive biopolitics' aimed at averting a future obesity and public health crisis (Kraftl 2015: 222).

Handy and Lee (2020) give an indication of how narrowing the current urban cycling agenda might be. In a study of cyclists in Davis, California (USA), Handy and Lee demonstrate that cycling does not adhere to the traditional logics of utility maximisation with regard to why people choose to cycle. Their study shows that enjoyment is an important quality that dictates whether people will cycle or not. They go on to break this down into eight categories, including activities, bonding, sensations and feelings, and sub-categories including adventure, exploration, speed, excitement, freedom and friends (2020: 3). Whilst the study also showed a gradual instrumentalisation of cyclists as they aged, it also showed that these qualities remained significant in their importance. Cycle infrastructures such as superhighways evidently prioritise qualities such as speed, but arguably do little to encourage other qualities deemed important in Handy and Lee's study to come to the fore. Rather, as the name suggests, superhighways narrow the potential experiences of cycling to encourage just those that can enhance productivity. Such narrowing sits at odds with the potential of cycling to engender a host of other qualities that enhance wellbeing and happiness.

Barr (2018) argues that the most positive outcome of restructuring transport systems lies 'in enabling us to reconnect with place-based communities through de-individualising the way we move, slowing down our movement and promoting "active travel"' (Barr 2018: 11). Barr goes on to note that this requires 'a revitalisation in promoting low carbon and low mobility places; in other words, to recognise the implicit connections between place and movement that exist' (Barr 2018: 13). Clearly for Barr, the slowing down that comes with cycling can allow other journey qualities to be foregrounded. As such, cycling (as Handy and Lee have shown) holds the potential to change our relationship with not only place, but also self, society and economy. Whilst it is unclear whether Barr sees the implementation of active travel infrastructure like superhighways as helping to realise this potential, I would argue that London's superhighways do little to foster the connections between place and movement that Barr (2018) highlights. Thus the prioritisation of superhighways represents a profoundly ontological politics because it reflects decisions about the kinds of realities we value and try to enact (Law and Urry 2004: 396). The evidence here suggests that current priorities reflect a very narrow, exclusive and partial valuation of what cycling is and should be.

Accordingly, in line with Kraftl (2014), we can theorise the drive to focus on commuter cyclists as the coming together of efficiency, sustainability and health agendas to produce citizens in waiting who are responsible for themselves and embody active commuting as an instrumental, healthy and efficient form of mobility. The production of such bio-power is key to reproducing a nation whose human capital is as productive and self-governing as possible (Kraftl 2014: 272).

Urban governance, networks and flows

However, we also need to ask the question, how did we get here? There is a need to critically assess 'the implications of this new logic in shaping the contours of the emblematic, exemplary and dominant sociotechnical-ecological fix for cities' (Hodson and Marvin 2009: 210). In particular, Hodson and Marvin raise questions about the roles of elites and particular interest groups in establishing what counts as feasible and desirable responses to issues like climate change and which cities and communities may be excluded through these processes (Hodson and Marvin 2009: 210, in Bulkeley 2010: 242).

Evidently we cannot explain the contemporary emphasis on urban cycle commuting as solely the result of 'state' actors or some kind of inevitable 'end-state'. As Hackworth (2007) has stated of Harvey's original notion of spatial fix in relation to the US Great Depression, 'it was much more than a simple policy event. The spatial fix was an intersection of capital, policy, and individual preferences' (95). Congruently, any explanation that seeks to link the two needs to take into account the involvement of both state and non-state actors in processes of neoliberalisation. If, as Mould argues, 'capitalism's creative drive is to either destroy alternatives or co-opt and transform them in to fertile ground for stable growth' (2018), we must acknowledge that cycling institutions and researchers from the UK's London Cycle Campaign, Cyclists' Touring Club (CTC) and the European Cyclists' Federation, and countless universities have been central to this endeavour. As this suggests, certain groups would appear to be calling the shots about what kinds of mobility count, and within this what kinds of cycling count – narrowing the possibilities of cycling through design and policy with the outcome that less productive forms of cycling are framed out.[11]

Here I wish to highlight the multiple actors, networks and flows that are producing the resurgence of cycling. As D'Andrea *et al.* (2011), among others, have stated, accelerating and increasingly complicated global flows (of ideas, objects, practices, people, etc.) over the past 30–40 years have led a number of scholars (including many cultural geographers) to study and problematise the relations between and within a diverse array of mobilities, particularly in the fields of migration, transport and tourism. In particular, as Beck *et al.* (2003) have demonstrated, knowledge systems, in second modernity, are characterised by shifting boundaries, loose foundations and cross-border movements. Cerny (2006) has emphasised a shift from concentrated political processes towards 'a more structurally differentiated, messy process of multi-level restructuring' (680). Indeed, as Cerny goes on to state, increasingly globalised knowledge economies are characterised by networked forms of governance where interventions in economic and social arenas take very different forms (Cerny 2006: 686). Fuchs and Hofkirchner (2005) go on to state that the way in which the acceleration of communicative mobility is outstripping that of physical mobility means that knowledge is increasingly 'quite substantially detached from territorial space', a process equally matched by its re-embedding and localisation elsewhere (254–255). Additionally, Bulkeley (2010) notes how a new wave of municipal networks mobilises private and non-governmental actors alongside the state (233). As Bulkeley goes on to suggest, we can no longer only consider municipal authorities but must also ask how other actors seek to govern through these networks (Bulkeley 2010: 233).

The transmission of cycling as an urban phenomena exemplifies these kinds of messy, mobile and multi-level forms of governance. To start with, the resurgence of cycle policy in London cannot be seen as purely the result of intentional state intervention; it is as much due to the ad hoc interventions of various local and international non-governmental and individual actors. As Stehlin (2014) has noted in the promotion of urban cycling in the US context:

> the currency of the various spatial forms of bicycle planning comes from the way certain cities, particularly in Scandinavia, are framed as desirable models of 'real' cycling urbanism. [...] idealized versions of cities like Copenhagen still capture the interest of participants in these communities.
>
> (Stehlin 2014: 28)

Cycling as a particular kind of fix is instituted not from within the city-state or city-region but rather by drawing upon a variety of global actors and sources deemed to be 'expert' in order to enhance the competitiveness of the city. Golbuff (2014), for example, has illustrated in detail how policy knowledge related to cycling in London is increasingly imported and transformed through practices such as blogging and policy tourism. Not only does this suggest that the policy flows that inform such fixes are more dispersed and networked, Golbuff also suggests that such networks are to some extent self-organising (because of technologies such as hyper-linking) and reactive to shortcomings in existing policy, reflecting a neoliberal emphasis on citizens as self-activating (Golbuff 2014: 132).

To give another example, the London Cycling Campaign (LCC) has been instrumental in making the case for cycling in London. Launched in 1978 and originally closely linked to the green movement (particularly Friends of the Earth), in its early days, the campaign was much more guerrilla-minded, as evidenced by its organisation of protest rides and impromptu taping of cycle-friendly road markings and signage to streets (Spinney 2010). However, this radical stance has been increasingly replaced with an arguably more diplomatic strategy focused on improving conditions within the existing legislative and operative framework. The LCC, for example, have been one of the key stakeholders for Transport for London when undertaking cycle route inspections and implementations (Spinney 2010). This suggests, as Lin (2009) has argued, that under neoliberal governance, 'local government, social groups and communities are asked to become responsible for transforming themselves as part of a process of involvement with the construction and operation of local governance' (Lin 2009: 14). Certainly it is clear that previously marginal cycling groups such as the LCC and Sustrans have become professionalised and central to policy debates on cycling.

The involvement of groups such as the LCC and individual bloggers demonstrates that it is not only what we would most obviously describe as 'state' actors, such as TfL, who have driven the agenda for cycling in London, but also non-state actors. The involvement and agendas of these groups would also suggest that as much as we can see cycling as a kind of fix, it is evidently not one that is fully strategised on the part of all actors, rather it is of the metaphorical kind as described by Jessop (2004: 13). Indeed, as Jessop goes on to say, such fixes are the result of numerous actors coming together with the result that disparate and compromised agendas become institutionalised (Jessop 2004: 13). What is apparent in such a loose assembly of actors is the ability of state actors to be directed, but also to harness and channel the narratives of selected stakeholders in providing a fix. Bulkeley (2010) for one has noted the importance of transitional networks in urban engagements with climate change (231).

At least three conclusions can be deduced from this: first, that any spatial fixes associated with cycling are a by-product and adjunct to its relative failure as a purely mobility fix. Evidently what emerges as a mobility fix operating at the level of the individual becomes spatial to some degree because the individual entrepreneur faces numerous barriers (most notably safety) in getting the enterprise off the ground (for example, see Cramshaw 2012). In responding to these barriers, policy actors are forced to institute a limited suite of spatial fixes in order to help individual enterprises succeed. Following Çalışkan and Callon (2009, 2010), this limited suite of spatial fixes can be considered part of a process of agencement that seeks to align actors and actants in order to bring certain outcomes and subjects into being.

A second conclusion is that the mobility fix of cycle promotion/cycle infrastructure also represents what Jessop (2004) has termed a metaphorical fix: a 'second best' solution characterised by its improvised and temporary nature that at best partially and provisionally seeks to overcome the contradictions and dilemmas inherent in capitalism and regularise accumulation in the long term (Jessop 2004: 13). In their piecemeal and fragmented nature, cycling's spatial fixes appear avowedly metaphorical. Moreover, metaphorical fixes, as Jessop (ibid.) goes on to say, are an outcome of the institutionalisation of disparate and compromised agendas and actors coming together (as is the case here) in the kind of ad hoc policy assemblage described by Braun (2014).

Third, as Brenner has argued:

> the conditions for competitiveness in contemporary capitalism are increasingly determined by international, transnational and translocal imperatives. The main scalar winners, however temporarily, in the search for a new spatio-temporal fix are urban regions, which possess a range of uniquely appropriate and significant structural competitive advantages with regard to coordinating strategies of dispersion and strategies of concentration.
>
> (Brenner 2004: 214, cited in Cerny 2006: 689; see also Hodson and Marvin 2009)

We see an example of this in the case of cycling in London, where the solution to a range of primarily economic problems is realised in the extraction of behavioural surplus in cycling as a realm of social reproduction activity.

Conclusions

For the reasons outlined above, the key contribution of this chapter is to theorise contemporary cycling practices in London (particularly commuter cycling) as a mode of neoliberal governance. This chapter has demonstrated that promotion of cycling has become a key way in which responsibilities are shifted from public to private spheres in relation to a bio-politics that quantifies, monetises and constructs matters of concern such as obesity and sedentarism as abnormal and problematic and presents cycling as a solution. In doing so I contend that cycle promotion in London represents a bio-political mobility fix that attempts to construct preferred bodily modes of comportment and social infrastructures to support them. At the same time the chapter has also shown that the initial success of cycle promotion in London has highlighted the shortcomings of the existing spatial infrastructures to accommodate these new forms of mobility, necessitating an (albeit minimal and geographically unequal and contingent) spatial/metaphorical fix in the form of cycle-specific infrastructure.

Within this, there are a number of themes worthy of emphasis. The first is that this processing of 'fixing' is dependent on a process of surplus extraction. To explore this, I have sought to borrow and adapt key concepts from Zuboff's (2019) work on surveillance capitalism. Here I have argued that beyond the value that accrues to individuals in conducting their commute by bicycle (e.g. time and money savings), the labour of the cyclist also produces a behavioural surplus that can be harnessed by the state because, directed towards multiple matters of concern, certain forms of mobility in certain contexts have greater labour power for a given labour time worked. The surplus generated through this is valorised by directly linking it to addressing specific urban problems, most notably the increasing sedentarism of human capital; the increased pollution generated by forms of movement reliant on fossil fuels; and the symbolic 'brand' value of cycling in presenting an image of the city as green and liveable. I argue here that it is primarily because of the additional value that can be extracted from the cycling subject (as opposed to, for example, the driving subject) that we see cities like London making concerted efforts to bring these subjects into being through discursive and subsequently spatial orderings. As such, forms of mobility such as cycling enable greater 'social productivity' than, for example, driving because of the net amount of value the cyclist can produce through their labour in excess of its own consumption requirements (Marx 1990: 293).

I also want to highlight that as a process of fixing, promotion of cycling engages with the individual in the realm of social reproduction. Where previous fixes aimed at enhancing human capital may have been directed toward the sphere of labour relations (such as reskilling of the workforce), when focused on producing cycle commuters, the mobility fix of cycling works at the level of social reproduction. In seeking to induce new skills, knowledge, and dispositions in citizens outside the sphere of work, the co-option of cycling represents part of a broader shift from a market economy to a market society, where ever-more aspects of life are brought within the realm of economic discipline. Rather than the journey to work being unpaid time that we can spend how we like, increasingly commute time is subject to bio-political moral differentiation and subsequent discipline, where some forms of conduct become more appropriate than others.

Following on from this, it is evident that the labour that constitutes certain forms of cycling is generative of greater surplus than others (and indeed that some forms of cycling might be antagonistic to this process of extraction). A prominent outcome of the linking of cycling with behavioural surplus through the facilitation of commuting cycling is that certain types of cycling are promoted as more productive and therefore important than others. That is not to say that cycling does not become meaningful to people in their everyday lives, rather that we see discursive framings and spatial arrangements come into being that make certain performances (like commuting) more likely to arise than other versions of cycling. As a result, variants of cycling that might be detrimental to valorising behavioural surplus (such as political activism, slow cycling, trick cycling, social cycling, leisure cycling, etc.) are discouraged. Accordingly, we also have to question the marginalisation of subaltern and alternative versions of cycling that might orient us differently towards the city and economy more broadly.

Boltanski and Chiapello (2018) argue that capitalism has become increasingly predatory, preying on the 'people, ideas, things and movements that are in direct opposition to it' (Mould 2018: 13). Mould argues that rather than outright rejection, neoliberal capitalism effectively stabilises these movements by naming them, offering them a place at the table, mainstreaming them and bringing them into the public consciousness, prepping them for commercialisation (ibid.). In so doing, it potentially leaches away alternative readings of these practices that do not fit the need for profit and any notion that they could be used to create alternative social worlds. Stehlin, for example, has commented that the focus on cycling commuting undercounts subaltern cycling practices; those of low-wage labourers navigating shift work and social services (Stehlin 2014: 31) or underemployed youth riding within estates and locales are often discounted. It is notable that forms of sport and leisure cycling are also exorcised from urban policy framings, particularly those deemed to be 'inefficient' or 'for kids', like BMX and mountain biking (Spinney 2008: 144).

Finally, I also draw attention to the implications for sustainability in continuing to prioritise forms of cycling that emphasise speed and the individual over other qualities. As Harvey (1991) has argued, capitalism exerts an imperative to accelerate turnover time beyond the 'social average' – e.g. the pace at which we might conduct our lives otherwise – thus promoting a trend toward faster and more mobility (Harvey 1991: 229). Whilst cycling has the potential to slow us down, through superhighways its promotion in London is geared towards accelerating mobility beyond the current social average (7.2mph in central London in 2018, according to TfL) to recuperate productivity and increase economic growth. In selectively linking residential areas with key areas of urban production and consumption, cycling is being utilised to reproduce existing (unsustainable) patterns of production and consumption rather than to slow movement down and enable a reflexive questioning of such goals and spatial arrangements. As Barr (2018) has noted, current policies do too little to question 'why we travel so much' or what the effects of such hypermobility might be for our societies (2).

As this suggests, it is not just how we transport ourselves, but the goals to which that transport is directed that dictate its sustainability. In the same way that increasing access times and greening journeys to airports through the provision of public transport is a questionable goal given the current climate crisis, so too is using cycling to speed up circulation times to maintain current patterns of production and consumption: whilst cycling may help to decouple transport from emissions, in its current prioritised variants in cities like London, it continues to feed growth rather than question. Even if the switch to a cycle journey does improve individual wellbeing, if that cycle journey helps to fuel continued and unsustainable economic growth, it ultimately contributes to a reduction in collective wellbeing. It is not only how we travel that matters, but the goals to which that travel is oriented that matter with regard to sustainability.

Notes

1 This chapter is based on a previously published paper: www.tandfonline.com/doi/abs/10.1080/24694452.2015.1124016
2 After a period of unabated road-building in the UK throughout the 1970s and 1980s, enthusiasm began to wane as the 'new realism' set in and it became apparent that building roads just generated more traffic, congestion and associated environmental and health issues (Goodwin *et al.* 1991, in Docherty and Shaw 2008: 8).
3 In late 2014 the UK government announced a funding package of £15 million for new roads schemes in England. However, all of the proposed schemes are for inter-urban rather than intra-urban areas. www.bbc.co.uk/news/business-30269231
4 It is also highly questionable whether any road-building programme would significantly speed up circulation times, in part because increasing city densities mitigate against car use in ways that suburban living did/does not.
5 Whilst a discourse dominates which positions public transport as sedentary, we might want to question the extent to which this is true. Webb (2014), for example, has demonstrated that those who use public transport are 25% less likely to be obese than those who don't (4).
6 It should be noted that the state in such a reading is increasingly the municipal government.
7 Despite the strong (some might say overly so) associations of cycling with environmental issues, this paper does not discuss them as a component of the mobility fix. This is partly due to word limits, but more that this paper views contemporary cycling practice as a feature of neoliberal urbanisation that is relatively independent of the sustainability imperative.
8 This is against a backdrop of continual decline in UK cycling levels, which peaked in 1949 at 37% of all traffic and have been on the slide ever since, particularly between 1950 and 1970, and flat-lining throughout the 1980s (Golbuff and Aldred 2011: 3). By 1989, an EU survey placed Britain alongside Belgium as the worst cycling nation in Europe (Golbuff and Aldred 2011: 9), with official interest in cycling returning only slowly in the early 1990s.
9 An HGV is any goods vehicle that exceeds 3.5 tonnes in gross weight. There are two categories in the UK: N2 for vehicles between 3.5 and 12 tonnes, and N3 for those over 12 tonnes.
10 London has invested in two kinds of route: high-capacity direct and separated 'superhighways', and less direct, piecemeal and often unseparated quietways on low-traffic side streets. The progress of quietways has been much slower, with only one opened by mid-2018. By July 2019, four quietways were operational.
11 Certainly one could argue that cycle activists have (understandably) been so excited that someone has at last taken cycling seriously, they haven't stopped to

question why only fast cycling and commute corridors have been prioritised, and more locally based and 'unproductive' cycling has been marginalised.

References

Aldred, R. (2015) A matter of utility? Rationalising cycling, cycling rationalities, *Mobilities*, 5(1), pp. 686–705.

Azimont, F. and Araujo, L. (2007) Category reviews as market-shaping events, *Industrial Marketing Management*, 36, pp. 849–860.

Baerenholdt, J.O. (2013) Governmobility: The powers of mobility, *Mobilities*, 8(1), pp. 20–34.

Barr, S. (2018) Personal mobility and climate change, *Advanced Review*, pp. 1–19.

Beck, U., Bonss, W. and Lau, C. (2003) The theory of reflexive modernization: Problematic, hypotheses and research programme, *Theory Culture Society*, 20(2), pp. 1–33.

Bohm, S., Jones, C., Land, C. and Paterson, M. (2006) Introduction: Impossibilities of automobility, *Sociological Review*, 54(s1), pp. 2–16.

Boltanski, L. and Chiapello, E. (2018) *The New Spirit of Capitalism*, trans. Gregory Elliott (London: Verso).

Bonham, J. (2006) Transport: Disciplining the body that travels, *Sociological Review*, 54(s1), pp. 57–74.

Braun, B. (2014) A new urban dispositif? Governing life in an age of climate change, *Environment and Planning D: Society and Space*, 32, pp. 49–64.

Brenner, N. (2004) *New State Spaces: Urban governance and the rescaling of statehood* (New York/Oxford: Oxford University Press).

Briginshaw, D. (2019) Crossrail project likely to open in 2021 and cost more, *Rail Journal*. Available at: www.railjournal.com/passenger/commuter-rail/crossrail-to-cost-more/ [accessed November 2019].

Bulkeley, H. (2010) Cities and the governing of climate change, *Annual Review of Environment and Resources*, 35, pp. 229–253.

Button, K. and Nijkamp, P. (1997) Social change and sustainable transport, *Journal of Transport Geography*, 5(3), pp. 215–218.

Çalışkan, K. and Callon, M. (2009) Economization, part 1: Shifting attention from the economy towards processes of economization, *Economy and Society*, 38(3), pp. 369–398.

Çalışkan, K. and Callon, M. (2010) Economization, part 2: A research programme for the study of markets, *Economy and Society*, 39(1), pp. 1–32.

Callon, M. (2007) An essay on the growing contribution of economic markets to the proliferation of the social, *Theory, Culture and Society*, 24(7–8), pp. 139–163.

Callon, M., Méadel, C. and Rabeharisoa, V. (2002) The economy of qualities, *Economy and Society*, 31(2), pp. 194–217.

Carstensen, T., Olafsson, A., Bech, N., Poulsen, T. and Zhao, C. (2015) The spatio-temporal development of Copenhagen's bicycle infrastructure 1912–2013, *Geografisk Tidsskrift-Danish Journal of Geography*, 115(2), pp. 142–156.

Cerny, P.G. (2006) Restructuring the state in a globalizing world: Capital accumulation, tangled hierarchies and the search for a new spatio-temporal fix, *Review of International Political Economy*, 13(4), pp. 679–695.

Chartered Institute of Highways and Transportation (1996) *Cycle-friendly Infrastructure: Guidelines for planning and design* (London: CIHT).

Cramshaw, P. (2012) Governing at a distance: Social marketing and the bio-politics of responsibility, *Social Science and Medicine*, 75, pp. 200–207.
Cresswell, T. and Uteng, T. (2008) *Gendered Mobilities* (Farnham: Ashgate).
Cupples, J. and Ridley, E. (2008) Towards a heterogeneous environmental responsibility: Sustainability and cycling fundamentalism, *AREA*, 40(2), pp. 254–264.
D'Andrea, A., Ciolfi, L. and Gray, B. (2011) Methodological challenges and innovations in mobilities research, *Mobilities*, 6(2), pp. 149–160.
Davis, A. (2010) *Value for Money: An economic assessment of investment in walking and cycling* (NHS Bristol).
Dean, M. (2010) *Governmentality: Power and rule in modern society*, 2nd ed. (London: Sage).
Department for Transport (2008) *Local Transport Note 2/08* (London: The Stationery Office).
Docherty, I. and Shaw, J. (2008) *Traffic Jam: Ten years of sustainable transport in the UK* (Bristol: Policy Press).
Doward, J. (2013) Westminster's world heritage status at risk as UNESCO condemns plans for skyscrapers, *The Guardian*, 9 June. Available at: www.theguardian.com/culture/2013/jun/09/westminster-world-heritage-status-risk [accessed 21 August 2014].
Entwistle, J. (2006) The cultural economy of fashion buying, *Current Sociology*, 54(5), pp. 704–724.
Fincham, B. (2006) Bicycle messengers and the road to freedom, *Sociological Review*, 54(s1), pp. 208–222.
Foucault, M. (2010) *The Birth of Bio-politics: Lectures at the college de France 1978–79* (Basingstoke, UK: Palgrave Macmillan).
Fuchs, C. and Hofkirchner, W. (2005) Self-organization, knowledge and responsibility, *Kybernetes*, 34(1/2), pp. 241–260.
Furness, Z. (2005) *Put the Fun between Your Legs! The politics and counterculture of the bicycle*, unpublished PhD thesis (University of Pittsburgh).
Garrard, J., Crawford, S. and Hakman, N. (2006) *Revolutions for Women: Increasing women's participation in cycling for recreation and transport: Summary of key findings* (Victoria, Australia: School of Health and Social Development, Deakin University).
Golbuff, L. (2014) *Moving Beyond Physical Mobility: Blogging about urban cycling and transport policy*, unpublished PhD thesis (Cardiff University, UK).
Golbuff, L. and Aldred, R. (2011) *Cycling Policy in the UK: A historical and thematic overview* (UEL Sustainable Mobilities Group, University of East London).
Golub, A., Hoffman, M. and Lugo, A. (eds) (2016) *Bicycle Justice and Urban Transformation: Biking for all?* (Abingdon, UK: Routledge).
Greater London Authority (2014) *Cycle Flows on the TFL Road Network* (London: GLA). Available at: http://data.london.gov.uk/dataset/cycle-flows-tfl-road-network [accessed 26 November 2014].
Greater London Authority Economics (2005) *Time is money: the economic effects of transport delays in central London* (Oxford Economic Forecasting, GLA London). Available at: www.london.gov.uk/mayor/economic_unit/docs/time_is_money.pdf [accessed 1 November 2014].
Greco, M. (2009) Thinking beyond polemics: Approaching the health society through Foucault, *Österreichische Zeitschrift Für Soziologie*, 34(2), pp. 13–27.
Guthman, J. (2009) Teaching the politics of obesity: Insights into neoliberal embodiment and contemporary biopolitics, *Antipode*, 41(5), pp. 1110–1133.

Hackworth, J. (2007) *The Neoliberal City: Governance, ideology and development in American urbanism* (Ithaca, NY: Cornell University Press).

Handy, S. and Lee, A. (2020) What is it about bicycling? Evidence from Davis, California, *Travel Behaviour and Society*, 20, pp. 348–357.

Harvey, D. (1982) *The Limits to Capital*, 2nd edition (London: Verso).

Harvey, D. (1989) From managerialism to entrepreneurialism: The transformation in urban governance in late capitalism, *Geografiska Annaler*, 71B(1), pp. 3–17.

Harvey, D. (1991) *The Condition of Post-Modernity: An enquiry into the origins of cultural change* (Malden, MA: Wiley-Blackwell).

Hickman, R. and Bannister, D. (2015) *Transport, Climate Change and the City* (London and New York: Routledge).

Hodson, M. and Marvin, S. (2009) 'Urban ecological security': A new urban paradigm?, *International Journal of Urban and Regional Research*, 33(1), pp. 193–215.

Hoffman, M. and Lugo, A. (2014) Who is 'world class'? Transportation justice and bicycle policy, *Urbanities*, 1, pp. 45–61.

Huxley, M. (2006) Spatial rationalities: Order, environment, evolution and government, *Social and Cultural Geography*, 7, pp. 771–787.

Jennings, G. (2015) Finding our balance: Considering the opportunities for public bicycle systems in Cape Town, South Africa, *Research in Transportation Business and Management*, 15, pp. 6–14.

Jessop, B. (2004) *Spatial Fixes, Temporal Fixes, and Spatio-Temporal Fixes*. Available at: http://bobjessop.org/2014/01/16/spatial-fixes-temporal-fixes-and-spatio-temporal-fixes/ [accessed 1 November 2014].

Jones, P. (2012) Sensory indiscipline and affect: A study of commuter cycling, *Social and Cultural Geography*, 13(6), pp. 645–658.

Jungnickel, K. and Aldred, R. (2014) Cycling's sensory strategies: How cyclists mediate their exposure to the urban environment, *Mobilities*, 9(2), pp. 238–255.

Kraftl, P. (2014) Liveability and urban architectures: Mol(ecul)ar biopower and the 'becoming lively' of sustainable communities, *Environment and Planning D: Society and Space*, 32, pp. 274–292.

Kraftl, P. (2015) Alter-childhoods: Biopolitics and childhoods in alternative education spaces, *Annals of the Association of American Geographers*, 105(1), pp. 219–237.

Krizek, K.J., Johnson, P.J. and Tilahun, N. (2005) *Gender Differences in Bicycling Behaviour and Facility Preferences: Research on women's issues in transportation*, Conference Proceedings 35, Technical Papers, vol. 2 (Washington, DC: Transportation Research Board).

Lam, T.F. (2018) Hackney: A cycling borough for whom?, *Applied Mobilities*, 3(2), pp. 115–132.

Law, J. and Urry, J. (2004) *Enacting the Social* (Lancaster: Department of Sociology and the Centre for Science Studies, Lancaster University). Available at: www.comp.lancs.ac.uk/sociology/papers/Law-Urry-Enacting-the-Social.pdf.

Lees, L. (2003) 'Visions of "urban renaissance": The Urban Task Force Report and the Urban White Paper', in Imrie, R. and Paco, M. (eds) *Urban Renaissance? New Labour, community and urban policy* (Bristol: Policy Press), pp. 61–82.

Leet, D.J., Ho, H., Banks, M., Giampieri, M., Chen, X. and Le, D. (2016) 'Delivering (in)justice: Food delivery cyclists in New York City', in Golub, A., Hoffman, M. and Lugo, A. (eds) (2016) *Bicycle Justice and Urban Transformation: Biking for all?* (London: Routledge), pp. 114–129.

Lin, W-I. (2009) *Urban Renaissance and Community Governance in London and Shanghai*, unpublished PhD thesis (Kings College London).
London Assembly (2017) *Mayor and Commissioner Set Out Vision for Getting Londoners Active*. Available at: www.london.gov.uk/press-releases/mayoral/setting-out-a-vision-for-getting-londoners-active [accessed August 2019].
Marx, K. (1990) *Capital, Volume 1*, trans. Ben Fowkes (London: Penguin Group).
McFall, L. (2009) Devices and desires: How useful is the 'new' new economic sociology for understanding market attachment, *Sociology Compass*, 3(2), pp. 267–282.
Meador, D. (2016) 'Cycling on the U.S./Mexican border with Doble Rueda bicycle collective in Matamoros, Tamaulipas', in Golub, A., Hoffman, M. and Lugo, A. (eds) *Bicycle Justice and Urban Transformation: Biking for all?* (London: Routledge), pp. 156–171.
Minn, M. (2013) The political economy of high speed rail in the United States, *Mobilities*, 8(2), pp. 185–200.
Mould, O. (2018) *Against Creativity* (London: Verso).
Nettleton, S. and Green J. (2014) Thinking about changing mobility practices: How a social practice approach can help, *Sociology of Health and Illness*, 36(2), pp. 239–251.
Newman, P., Kenworthy, J. and Vintila, P. (1995) Can we overcome automobile dependence? Physical planning in an age of cynicism, *Cities*, 12(1), pp. 53–65.
Nielsen. T.A., Skov-Petersen, H. and Carstensen, T.A. (2013) Urban planning practices for bikeable cities – the case of Copenhagen, *Urban Research and Practice*, 6(1), pp. 110–115.
Nixon, D. (2012) A sense of momentum: Mobility practices and dis/embodied landscapes of energy use, *Environment and Planning A*, 44(7), pp. 1661–1678.
Organisation for Economic Cooperation and Development (2014) *The Cost of Air Pollution: Health impacts of road transport* (Paris: OECD Publishing).
Office for National Statistics (2014) *London Residents Cycling to Work Doubles in 10 Years* (London: ONS). Available at: www.ons.gov.uk/ons/rel/census/2011-census-analysis/cycling-to-work/sty-cycling-to-work.html [accessed 30 November 2014].
Pooley, C., Horton, D., Scheldeman, G., Mullen, C., Jones, T., Tight, M., Jopson, A. and Chisholm, A. (2013) Policies for promoting walking and cycling in England: A view from the street, *Transport Policy*, pp. 66–72.
Robertson, A. (2001) Biotechnology, political rationality and discourses on health risk, *Health: An Interdisciplinary Journal of Social Studies in Health and Medicine*, 5(3), pp. 293–309.
Rose, N. (1999) *Powers of Freedom: Reframing political thought* (Cambridge: Cambridge University Press).
Spinney, J. (2006) A place of sense: A kinaesthetic ethnography of cyclists on Mt Ventoux, *Environment and Planning D: Society and Space*, 24(5), pp. 709–732.
Spinney, J. (2007) 'Cycling the city: Non-place and the sensory construction of meaning in a mobile practice', in Horton, D., Rosen, P. and Cox, P. (eds) *Cycling and Society* (Aldershot, UK: Ashgate), pp. 25–46.
Spinney, J. (2008) *Cycling the City: Movement, meaning and practice*, unpublished PhD thesis (Royal Holloway University of London).
Spinney, J. (2010) 'Mobilising sustainability: Partnership working between a pro-cycling NGO and local government in London (UK)', in Fudge, S., Peters, M. and Jackson, T. (eds) *Low Carbon Communities: Imaginative approaches to combating climate change* (Cheltenham, UK: Edward Elgar), pp. 89–107.

Spinney, J., Kullman, K. and Golbuff, L. (2015) Driving the 'Starship Enterprise' through London: Constructing the im/moral driver-citizen through HGV safety technology, *Geoforum*, 64, pp. 333–341.

SQW Consulting (2008) *Planning for Cycling: Report to Cycling England*. Available at: http://webarchive.nationalarchives.gov.uk/20110407094607, http:/www.dft.gov.uk/cyclingengland/site/wp-content/uploads/2009/03/planning-for-cycling-report-10-3-09.pdf [accessed 1 January 2014].

Stehlin, J. (2014) Regulating inclusion: Spatial form, social process, and the normalization of cycling practice in the USA, *Mobilities*, 9(1), pp. 21–41.

Stehlin, J. (2019) *Cyclescapes of the Unequal City: Bicycle infrastructure and uneven development* (Minneapolis: University of Minnesota Press).

Tickell, A. and Peck, J. (2003) 'Making global rules: Globalization or neoliberalization?', in Peck, J. and Yeung, H. (eds) *Remaking the Global Economy: Economic-geographical perspectives* (London: Sage), pp. 163–181.

Transport for London (2010) *Cycling Revolution London* (London: TfL).

Transport for London (2013) *Annual Report and Statement of Accounts* (London: TfL). Available at: www.tfl.gov.uk/cdn/static/cms/documents/annual-report-2013-14.pdf [accessed 7 November 2014].

Transport for London (2014) *London's Road Modernisation Plan* (London: TfL).

Transport for London (2018) *15km of New Quietway Cycle Routes Opened across London*. Available at: https://tfl.gov.uk/info-for/media/press-releases/2018/october/15km-of-new-quietway-cycle-routes-opened-across-london [accessed November 2019].

Transport for London (2018) *Freedom of Information Request Detail: Quietways Funding*. Available at: https://tfl.gov.uk/corporate/transparency/freedom-of-information/foi-request-detail?referenceId=FOI-3001-1718 [accessed August 2019].

Twaddle, H., Hall, F. and Bracic, B. (2010) Latent bicycle commuting demand and effects of gender on commuter cycling and accident rates, *Transportation Research Record*, 2190(1), pp. 28–36.

Webb, E. (2014) *Bus Passes, Active Travel and Health*, ICLS Occasional Paper, 11(3) (London: University College London).

While, A., Jonas, A.E.G. and Gibbs, D.C. (2004) The environment and the entrepreneurial city: Securing a 'sustainability fix' in Leeds and Manchester, *International Journal of Urban and Regional Research*, 28, pp. 549–569.

White, P. (2008) 'Transport for London: success despite Westminster', in Docherty, I. and Shaw, J. (eds) *Traffic Jam: Ten years of sustainable transport in the UK* (Bristol: Policy Press), pp. 183–204.

Whitehead, M., Jones, R., Lilley, R., Pykett, J. and Howell, R. (2017) *Behavioural Government in the 21st Century* (London: Routledge).

Zuboff, S. (2019) *Surveillance Capitalism: The fight for a human future at the new frontier of power* (New York: Hachette Book Group).

5 Economising 'trick' cycling on London's South Bank[1]

Culture-led regeneration, spectacle and 'entertailing'

Introduction

> It's about 10am on a Saturday morning and I'm up at the Shell Centre near Waterloo watching about 10 riders jumping and hopping on and off the street furniture on slightly strange looking bikes. A couple of the riders are sitting on a bench changing a flat tyre, the rest are doing an assortment of seemingly impossible moves using a variety of street ledges and drop-offs, all on the front or back wheel, but rarely both. After a few minutes some of the riders move across the road and 'attack' a wall in front of the London Eye. It's an almost vertical 4ft ramp with another smaller 'box' section raised another 3ft. They take turns in riding across the pavement and attempting to hop up on the ledge from a variety of different points. Half an hour later everyone drifts into a disheveled car park between the Eye and the Royal Festival Hall. There are stacks of precariously balanced and damp building supplies here and the riders straight away create a challenge out of them. It's an incredible display of skill and balance as they wend their way round obstacles, making their route up on the spot depending upon what looks possible or challenging enough.
>
> (field diary, 11 September 2005)

Whilst these Trials[2] riders use a variety of London's spaces, like many BMX[3] riders and skateboarders they congregate regularly around London's South Bank. Like many waterfront areas across the world, the South Bank is currently being redeveloped with an emphasis on imagery and spectacle as part of a retail-led development (Dovey 2005: 12; Stevens 2007b). Thus the choice of location may seem quite conventional in some senses as the riders' route often takes in the scenic backdrop of the London Eye, the Tate Modern and St Paul's, to name a few landmarks. Despite these obvious visual attractions, these riders are on one level more concerned with the kinaesthetic potential of the local architecture (Bissell 2008; Borden 2001; Lorimer 2005; Spinney 2006, 2007). The unconventional use of these spaces by the riders would appear to be at odds with the mobilities that are more commonly normalised as 'appropriate' in a pedestrian public space such as the South Bank (Imrie and Street 2009; Lees 2004; Middleton 2009; Spinney 2008). Certainly many previous accounts of similar bodily practices such as skateboarding have consistently highlighted their resistant quality because of the way they transgress dominant architectural and management codes (Beal 1995; Borden 2001; Flusty 2000; Vivoni 2009). Certainly in many central public spaces, riders have faced a politics of mobility with much in common to that experienced by the homeless (May 2000; Cloke *et al.* 2010) and are frequently excluded from these areas by design or moved on (Grafham 2002).

One of the reasons for this relative distaste for such modes of movement is that they are deemed to be playful and destructive rather than instrumental and productive. Accordingly they do not fit in with the dominant framing of cycling in London outlined in Chapter 4. Lorimer and Marshall (2015) comment upon the reduction in current policy of open-ended imaginaries of active travel to walking and (certain forms of) cycling, arguing instead for a broader idea of 'human locomotion', which they suggest could also include consideration of 'travel by small-wheel modes: inline skates, skateboards and push scooters' (1) as well as other kinds of cycle, and which have the potential to help create more 'agile' cities. As they go on to note, 'young people predominate in the use of alternative and non-traditional modes of transport' and as a result these modes represent a 'new and innovative way of encouraging active travel among the young' (2). Accordingly I argue that these other(ed) forms of cycling and mobility need to be taken seriously in the context of liveable and age-friendly cities.

Whilst not disagreeing that practices such as skateboarding, Trials and BMX are deemed inappropriate in many public spaces, in this chapter I contest accounts which regularly result in a binary diagnosis of such practices being 'in place' or 'out of place' (Cresswell 1996). Instead I build upon the idea of the mobility fix outlined in previous chapters to illustrate how more playful cycling practices have become economised in service to broader economic goals. Whilst the managers of the South Bank may (at least initially) stop short of actively inviting spectacular practices like Trials and BMX riding, I argue that their management of them (tending toward toleration) and subsequent spatial accommodation of them downplays resistant qualities and emphasises qualities of spectacle and dynamism. By encouraging wheeled practices that exhibit these qualities, the curators of the South Bank seek to attract and keep people in these spaces, by overlaying meanings of place with 'culture' that deflect from its designation as a space of retail capital. Certainly, as the retail literature is keen to point out, spectacle in retail spaces is a key strategy used to maximise footfall and revenue. Accordingly, in this chapter I argue that through social and spatial management, specific qualities of mobilities such as Trials, BMX and skateboarding are emphasised and economised to produce a behavioural surplus that is harnessed in service to the logic of capital accumulation underpinning the remaking of waterfront retail spaces.

As this argument suggests, these mobile practices are about much more than bodies, architecture and subculture. I emphasise the importance of both social and affective *elements* of practice, management responses to it, and spatial framing in the process of economising these performances. In doing so I reject accounts that ostensibly see such practices as resistant, or that just see them as solipsistic. Certainly – and considering the South Bank's stated emphasis on encouraging spectacle and social encounters – I would argue that the interactions between Trials, BMX riders and other users of the space performs the South Bank in ways which actively nurture elements of these styles of riding that complement what the retail-led redevelopment is trying to achieve.

A portion of the empirical data within this chapter is drawn from a paper first published as Spinney (2010). This data was part of a wider study on urban cycling conducted between 2004 and 2008 (Spinney 2008). In revisiting and updating it here, as well as revising the theoretical framing of the data, I have also drawn upon more recent policy documents and news articles to bring the story of the South Bank up to date. The first half of the chapter is split into two parts: the first section briefly introduces the South Bank (and in particular the Shell Centre) and the key tenets of its redevelopment, positioning it alongside other similar waterfront and retail-led regeneration projects. In particular I note the importance of dynamism, experience and spectacle to the success of such developments. The second section moves on to critically discuss resistance in the context of different spatial practices including walking, skateboarding and cycling. Shackled to an overemphasis on body–architecture relations, subculture and specific research methods, I note a tendency to overemphasise resistance as a quality in relation to a pre-existing spatial hegemony. I then go on to discuss processes of economisation, arguing that through a gradual agencement of management and spatial accommodation on the South Bank, dynamic and spectacular qualities of Trials biking and BMX (and of course skateboarding) constitute a behavioural surplus that enhances the retail offering. In doing so, I argue that whilst previous strategies have often sought to move riders and skaters out of unofficial public spaces to peripheral purpose-built facilities premised upon an essentialised and misdiagnosed reading of resistance, this account suggests that such practices may increasingly be managed into specific urban spaces to (implicitly) enhance economic competitiveness.

London's South Bank: retail, regeneration and urban dynamism

London's South Bank is Europe's largest arts complex and began to emerge from 1951 onwards as part of the Festival of Britain, beginning with the opening of the Royal Festival Hall. Its initial conception was as a home for culture (Jones 2016), and this emphasis has continued with numerous other structures added in the following decades, including the Queen Elizabeth Hall in 1967, the Hayward Gallery in 1968 and the National Theatre in 1976. As Jones (2016) comments, these represent a commitment to a broadly civic culture which became more arts-led from the late 1980s (246). However, the architecture – notably the Hayward Gallery and the Queen Elizabeth Hall – has been derided for its 'concrete brutalism'. Cultural commentators have in recent years described the area as 'fast descending into squalor and dereliction' with 'huge dusty areas untouched for decades and eaten away by decay' (Sudjic 2001); and as an 'area only muggers could enjoy and where just finding an entrance door is a feat' (Blume 2001).

Consequently, the South Bank and its environs have been subject to ongoing major renovations beginning with the Coin Street end of the South Bank in the 1980s and 1990s (Brindley *et al.* 1996). A more recent phase of development begun in 1999 has seen the area around the Hayward Gallery, Royal Festival Hall and the Riverside made over with new walkways, shops, restaurants and frontages (South Bank Centre 2007) similar in intent to that of other waterfront redevelopments (Stevens and Dovey 2004: 351; see also Dovey 2005; Kearns and Philo 1993; Marshall 2001). Whilst there are numerous local authority and mayoral proposals to revitalise the area around the South Bank and Waterloo, Rick Mather's 'masterplan' for the area around the South Bank complex intends to address 'the wider urban issues present in the area to encourage greater activity, diversity and accessibility for the site as a whole' (Virginia Museum of Fine Arts n.d.). The central aims of Mather's plan are to bring the city into the area because, as he says, 'streets are wonderful things, the more lively and intense they are the better' (Mather, quoted in Blume 2001).

Inner city malls 'without malls'

Such redevelopment of London's South Bank is indicative of a wider trend in central urban planning narratives of the last 30 years to reclaim disused and rundown river frontages and reimagine them as vibrant consumer spaces (Dovey 2005: 9). This phenomenon emerged in the 1980s (Kearns and Philo 1993) when many urban development corporations (UDCs) enabled retailers to buy contaminated and rundown brownfield sites, and in doing so gain exemption to certain planning rules in order to kick-start urban regeneration (Wrigley and Lowe 2002). Writing in 2005, Lowe uses the example of West Quay shopping Mall in Southampton (UK) to suggest that such redevelopments represent a pioneering integration of retail development and regeneration (2005: 449). Whilst much of this redevelopment has focused on out-of-town and mall spaces, from the 1990s onwards, tightened planning requirements (due to revised tests in Planning Policy Guidance Note 6 – PPG6) and a focus on place-making heralded a focus on inner-city retail and culture-led development (Lowe 2005: 453). Boland (2013), for example, has discussed Liverpool's signature 'Liverpool One' retail development, a £920 million development consisting of shops, hotels, apartments, cafes, bars, restaurants, office space, and parkland, offering an 'open street system as opposed to the traditional mall-style development, [that] reconnects the city centre and waterfront' (Biddulph 2011: 85–90, in Boland 2013: 259).

Waterfronts have long been a staple of culture-led retail development, having much in common with a new phase of retail development that Kern (2008) identifies as 'malls without walls' (105): 'the reconfiguration of the fortress-like, interior-oriented mall into an open, outdoor simulated "Main Street", complete with sidewalks, street furniture, street parking and crosswalks' (Kern 2008: 105). Growing in popularity from the early 2000s, Kern argues that such spaces are 'conceived to supply urbane "experiences" for sophisticated wealthy shoppers, but in a hermetically safe "public space" – safe from the intrusion of real-world nuisances such as panhandlers, homeless people or street kids' (ibid.). Similarly to the 'lifestyle centres' that Kern (2008) focuses upon, waterfront retail relies on high-end retail and restaurants; architectural embellishments; distinctive exterior features and finishes; entertainment facilities and live events (including live music, markets and street performers) in order to recreate 'the amenities and ambience of a thriving urban street' and make them attractive destinations for more than just shopping (Kern 2008: 107).

Spectacle and retail experience

As Boland has argued, in the current climate, the dynamics of urban competition rely less on heritage, reputation and natural resource and more on the ability to create exciting places that interpellate culture, retail, entertainment and leisure (Boland 2013: 252). As Arnold and Reynolds (2003) point out, any focus on core retail strategies is redundant at a time when retailers need to focus on 'entertailing' and hedonic experience as a key competitive tool (77). As they go on to state, 'hedonic shopping motives are similar to the task orientation of utilitarian shopping motives, only the "task" is concerned with hedonic fulfillment, such as experiencing fun, amusement, fantasy, and sensory stimulation' (Arnold and Reynolds 2003: 78). The lineage of this emphasis on experience, entertainment and hedonism can be traced back to the early department stores and their denomination as 'Cathedrals of Consumption' (Ritzer 1999). Such spaces are designed 'with the anticipation that consumers will gradually develop a passion for acquiring goods and services after experiencing a spectacle … defined as a form of drama, a form of theater, and a craze of information flows, music, entertainment, and other strategies intended to attract consumers' (Matusitz 2012: 1).

The means of engineering such affects are often within the individual retail store, with face painting, rock climbing, golfing, boat rides, and off-roading all examples of activities that have been used by retailers to shape atmosphere and experience (Arnold and Reynolds 2003), and disguise the act of consumption in the spectacle. However, beyond these tightly choreographed events, the surrounding area and a more diffuse set of activities evidently also have a role to play in shaping how we experience retail spaces. Accordingly Dovey notes that waterfront redevelopments increasingly focus on imagery and marketing that position the location alongside spectacles of 'artistic, social and economic dynamism' (2005: 13). An industry report on the Branson Landing retail development on Lake Taneycomo in the US positions the development of its spectacular 'Grand Finale' water show explicitly in relation to enhancing the retail offering, stating that:

> the show starts out on the lake but it moves back into the main fountain of the town square … leaps over the spectators and draws the attention back into the centre of activity, returning the focus to the shops and restaurants surrounding the square.
>
> (Chain Store Age 2003: 160)

Evidently – and as Miller (2013) argues – these spaces attempt to engineer sales and society through 'intervention in the affective "atmosphere" of the retail environment itself' (845). In such instances, retail professionals and interest groups do not manipulate the qualities of individual products directly to induce consumption (Callon *et al.* 2002: 200), but rather seek to shape the qualities of the space and atmosphere in which consumption takes place.

Of specific interest is the focus on particular spatial practices within such urban imagineering, which, according to Dovey and Stevens (2005), serve to simulate and choreograph playful urban life for passive consumption, channelling spontaneity and spectacle into a 'contrived form of permanent carnival' (67). Indeed, the importance of urban life to the success of retail developments is highlighted in the case of West Quay in Southampton, where the council's stated vision was that 'it will be focused around a new open plaza capable of having special events, but animated by people and activity at all times of the day, week and seasons' (Southampton City Council 2001, in Lowe 2005: 463). Indeed, the activities in the broader retail space provide the 'mood music' and are capable of setting the dynamic, fun and adventurous tone that modern entertailing seeks. In line with this, according to Jones (2016), one of the key functions of the South Bank currently being promoted by the South Bank Employers Group (SBEG) is: 'The animation and orchestration of public space. An explicit will to exploit public realm for consumer ends – to assert the "exchange value" … of that space' (250).

Alternative mobilities, resistance and 'appropriate dynamism'

It would be easy to critique the makeover of the Shell Centre and the South Bank more generally for its apparent denial of alternative uses and homogenisation – part of the middle-class 'cappuccino culture' that has been well documented as being the prevailing form of post-modern urban regeneration. According to Stevens (2007b), since the late 1970s large areas of public space have been restructured to provide leisure landscapes that serve power and profit at the expense of particular forms of play and expression (52). Zukin notes how such redevelopments often erode the two core principles of public space – public stewardship and open access (Zukin 1995: 143; see also Belina and Helms 2003, on the policing of public space as a part of city image). Pinder argues that in such spaces, 'utopia is cast in terms of the market and global capitalism – an ideal space of free exchange and consumer satisfaction, running smoothly with flows of money and commercialised desire' (Pinder 2002: 236; see also Fyfe 1998). Similarly, Stevens notes that although the South Bank has many 'programmed' leisure activities, these are defined within a narrow range such as 'viewing' the waterfront (Stevens 2007b: 143; see also Lees 1998: 237; Davis 1992, 1998; Smith 2002). In such readings, whilst there is a rhetoric of democracy, control of space and the erasure of difference is more than ever on the agenda. Certainly, as Jones points out in the case of the South Bank, 'in accounts of those responsible for transforming the South Bank Centre there is an evident will not only to physically reshape the Centre's urban realm but also to *curate the content of that realm*' (Jones 2016: 246).

As Cresswell (2006) notes, particular groups have always had more power to define who should be able to move around and how they can do it. Thus numerous critics (Cosgrove 1996; Harvey 1989; Hill 1998, 2003; MacLeod and Ward 2002; Pinder 2002; Rendell 1998; Robins 1991; Zukin 1993, 1995) have argued that planners and architects possess the agency and power to define the use of space. Certainly, as MacLeod and Ward (2002) point out, commentators such as Zukin (1993, 1995) and Harvey (1989) appear to suggest that users quite passively receive prescribed uses of space as they were intended; staying out of them or dutifully performing the correct behaviours in them (163). Cresswell (1993), on the other hand, points out that although there is a dominant culture that works to manipulate and repress ideas and imagination, it is not all-powerful (250), and there are many practices that fall between the cracks (see also Pinder 2002; Robins 1991). Similarly, Dovey and Stevens (2005) caution against such closed readings of space, suggesting in the case of waterfront developments that they also 'create conditions for spontaneous, unexpected and resistant activities' (68).

Despite this, a key feature of waterfront retail developments is that of ensuring social order and homogeneity in the form of 'safe adventure' for 'responsibilised/respectful citizens' (Boland 2013: 260). The huge shopping and leisure complex of Liverpool One, for example, is a 'privatised urban space', as its redevelopment entailed granting a 250-year lease to the developer. As a consequence, similarly to other mall developments (Staehili and Mitchell 2006), it is now policed by private security with powers 'to eject those not behaving in a proper manner or deemed to be potentially troublesome' (Boland 2013: 261). Such management practices suggest a strong desire to erase qualities of place/activity that may put consumers off (e.g. danger, discomfort) and emphasise those that induce consumption (e.g. safety, comfort, spectacle). These tendencies invite us to question exactly what are desirable carnivalesque activities in such spaces. The literature frequently flags up that activities like begging, homelessness, political activism and gang activity are strictly policed out of such environments (Kern 2008; Staehili and Mitchell 2006). Activities such as skateboarding, Parkour, BMX and Trials biking have a more complex relationship and seem to be more or less tolerated depending on the development's management and what they consider to be appropriate spectacle (Jones 2016; Stevens 2007a, 2007b; Spinney 2008).

The history of cycling as a form of spectacle may give some clues as to why it may be more accepted as part of street performance. According to Darley (2000):

> spectacle … exists for itself (and by extension the other senses). Drained of meaning, bereft of the weight of fictional progress, the cunning of spectacle is that it begins and ends with its own artifice; as such, spectacle is simultaneously both display and on display.
>
> (Darley 2000: 104)

The use of the bicycle in a spectacular performance is hardly new; since the bicycle was invented, riders have been using it to perform. The world-famous Kaufmann Troupes, for example, were one of many trick cycling acts in a competitive industry and were booked as part of variety acts, circuses and Vaudeville up until at least 1907 (Markham 1996: 69–71). Such performances have done and still do largely revolve around balance: 'as riding a velocipede or high-wheeler depended on the rider's ability to maintain equilibrium, most performances involved balancing a bicycle when in motion or remaining still while performing some acrobatic feat' (Dodge 1996: 149). In the early days of cycling, many performances took place in the context of a circus or other organised event, although as Dodge again points out, this was not always the case, with many performances taking place 'wherever a rider could find a crowd – at fairs, gymnasiums, music halls, circuses, parades, races or club meets' (1996: 149). Echoing these early days, contemporary performances of Trials and BMX riders in public spaces can be seen as a continuation and evolution of this style of riding as spectacle. Accordingly they may not be as 'resistant' to the ideals of modern retail waterfront spaces as might first be imagined.

A key argument made in this chapter is that whilst the South Bank Business Improvement District (BID) and the private security 'South Bank Patrol' it funds (South Bank BID 2019: 4) could exclude Trials and BMX riders (and certainly works to exclude unspectacular forms of transgression), managing these mobilities and allowing them in certain areas enables qualities of dynamism and spectacle to be foregrounded. Valorising such aspects of the 'work' of Trials and BMX riders produces a behavioural surplus utilised to add value in distinguishing the South Bank from other retail spaces in its liveliness. This logic fits with the shift identified by Jones (2016) from open space to curated space in the external areas of the South Bank (251). Jones goes on to state that through the selection of permitted activities and events, the South Bank is 'arguably once again produced as "exhibition", as "a temporary space for public presentation"' (ibid.). My argument here is that a part of that exhibition includes Trials, BMX and skateboarders who – as we shall see – have moved from being tolerated to formally curated within this vision. Contrary to accounts that foreground resistance[4] as a key quality in any unsanctioned practices, in the following sections I illustrate the ways in which both structures and practices are performed.

Economising embodied practice

Cycling – and Trials and BMX specifically – are social practices, so the question becomes how are these social practices of cycling rendered economic in the context of the South Bank? Mobility is not just magically converted into other forms of capital or value, so how does this process of 'economisation' (Çalışkan and Callon 2009, 2010) occur? Social life does not just become economically relevant; dynamism and spectacle from an excess of playful urban life do not just become an economic resource but must be nurtured. Congruently I wish to elaborate more specifically on the mechanisms through which social practice is rendered economic. In order to do so I draw heavily upon the work of Callon (alongside numerous collaborators) and Foucault.

As I elaborate in the course of the chapter, the crux of the issue is the matter of concern already outlined that seeks to regenerate the South Bank and in so doing identify it as both cultural and retail space. In aspiring to this new identity, the South Bank must produce and accommodate new social identities, 'whose appearance is related to the expression of such matters of concern' (Callon 2007: 146). These social identities can be more or less coherently strategized, and in the case of the South Bank can refer to arts and cultural venues, events, retailers, hospitality, attractions, forms of mobility, activities, etc. Thus what Marilyn Strathern (1999) refers to as the 'proliferation of the social' can be identified as a direct result of the functioning of markets (Callon 2007: 146). However, in order to realise its aims, the range of social identities accommodated on the South Bank must be carefully managed to align with its goals.

Much of the framework that Zuboff (2019) uses to describe the logic of surveillance capitalism is pertinent here. Whilst mobility is the raw material necessary for the manufacture of behavioural surplus, extraction is premised upon particular 'social relations and material infrastructure with which the firm[5] asserts authority over those raw materials' (Zuboff 2019: 65).

Whilst there is no obvious 'firm' in the context of the South Bank, this is important because it suggests that the ways in which mobility becomes accepted and accommodated (or not) define whether surplus can be extracted or not.

Thus the ways in which particular movements are allowed and normalised becomes an important foci of analysis because of the ways in which it nurtures mobility in directions amenable to the extraction of surplus.

Foucault (1995: 163–169) asks how the human body becomes docile and useful in the context of a particular social system? Foucault invites us to ask why certain subjectivities are legitimised as normal and desirable whilst others are marginalised or excluded. In the context of the South Bank redevelopment, the desirable subject is one who can contribute to the realisation of a dynamic urban retail environment. Hence what we see, according to Foucault, is an extension of market rationality to domains that are not primarily economic (2010: 323) through an anatomo-politics that seeks to produce and accommodate certain forms of comportment and not others.

What I aim to demonstrate in the following section is that Trials and BMX riding on the South Bank are economised through a diffuse process of 'agencement': a socio-technical 'arrangement of heterogeneous constituents that deploys ... rules and conventions; technical devices ... logistical infrastructures; texts, discourses and narratives ... as well as the competencies and skills embodied in living beings' (Çalışkan and Callon 2010: 3). This process revolves around the coming together of riders, spectators, security staff, built environments, and social norms around public space and their contingent adjustment to each other, both in place and over time, so that they can ultimately be utilised for gain in this setting. This agencement is premised upon recognition that such practices can be useful, and an ongoing process of adoption and adaptation so that the economically 'useful' aspects are retained whilst more 'anti-social' elements are discouraged. Such agencement, then, is premised upon the attempted management of undesirable and desirable qualities. Central to this process is the fact that economic value is released through 'accumulation by dispossession' 'at very low (and in some instances zero) cost' (Harvey 2005: 153, in Zuboff 2019: 99). Unlike officially sanctioned 'events', the activities of Trials and BMX riders do not need to be booked or paid for, neither do they need their own venue in order to perform, so the cost of facilitating them can be little or nothing.

Street performances: making a scene

The Shell Centre and its courtyard are situated at the entrance to the South Bank promenade between Waterloo station and the river front, adjacent to Jubilee Gardens (see Figure 5.1). The Waterloo Open Space (WOS) steering group report *Space for All* suggests that the area (created in 1953) 'allows a visual link to the London Eye for pedestrians travelling to and from the River and Waterloo Station' (Farrer Huxley Associates 2004: 109).

Figure 5.1 The Shell Centre at London's South Bank.

Source: Author.

That said, it's not immediately clear exactly what the courtyard[6] surrounding the Shell Centre is for, but it would appear that it is designed to facilitate the movement of pedestrians from the main street to the waterfront with little distraction. The idea that this area is designed as a thoroughfare is backed up by the WOS report, which states that 'there is currently no [street] furniture but people use steps etc. for seating' (Farrer Huxley Associates 2004: 109), and that 'the site functions as a pedestrian route and sitting place, and acts as a stepping stone, linking Waterloo Station with Jubilee Gardens' (136).

Clearly such an area is not intended for use by BMX and Trials riders (or indeed any other cyclists[7]); pedestrian bodily practices are given prominence in the make-over of the area, as the Lambeth Council *Waterloo Area Supplementary Planning Document* (WSPD) notes: 'participants wanted to see street and public realm facilities geared towards pedestrians' (Lambeth Council 2009: 9). The only mention of cycling in any of the documents relating to the area's redevelopment is in relation to cycle commuting to improve safety, parking and access. As a result, street furniture designed in accordance with the South Bank Employer's Group (SBEG)[8] *Streetscape Design Guide* should 'be located in a position that does not impede pedestrian flows or pedestrian desire lines' and 'designed to discourage anti-social behaviour such as rough sleeping and skateboarding' (Lambeth Council 2009: 74). The courtyard itself contains some sculpture and raised planters, giving an architecture made up of steps, balustrades, blocks and hand rails – exactly the kind of area that BMX, Trials riders and skaters would find attractive. Congruently, the WOS report notes that 'the site is popular with skateboarders' (Farrer Huxley Associates 2004: 109).

As this suggests, the design codes embedded in the Shell Centre and South Bank more generally are attractive to BMX and Trials riders, a key reason why they often meet there in groups. Stevens (2007b) argues that various architectural and geographical features serve to attract different forms of unsanctioned leisure in such commodified spaces. Franck and Stevens (2007) point out these are often re-appropriated to serve other ends:

> walls, fences and ledges, which are often supposed to delimit space and behaviour, can be sat upon, climbed onto and used to display banners or items for sale; their looseness is a product of affordances which such boundaries provide.
>
> (Franck and Stevens 2007: 8)

The Trials riders make use of architectural features such as rails, planters, stairs, edges, gaps, and ledges in order to perform any number of explosive and balancing manoeuvres. Rather than deterring riders, these features are actively sought out by riders:

> It's a very wet morning and I'm at the Shell Centre in central London watching and chatting to some Trials riders who are busy moving around the street furniture and chatting. Some of them are using the low ledge in the centre of the area as well as the stairs, some odd-shaped railings and a 5 ft ledge which I've no idea how they're getting onto. One guy is about to attempt this and I point the camera his way. He prepares himself carefully and takes a 10 ft run up to launching himself off the ground with the aim of jumping the 5 ft to get his back wheel onto the ledge. He narrowly misses, with his back wheel slipping off so that

> he slips back and smacks the tiny crank and bash-guard on the ledge. Unfazed, he simply goes back and tries again. And again. And again.
>
> (field diary, 11 September 2005)

These observations foreground the use of challenging walls, vertices and gaps, features that are reimagined when approached from the perspective of the hybrid bike-rider. In his study of the area, Stevens (2007b) also noted such uses at South Bank, affirming that physical edges are used by riders because they afford 'sudden, thrilling exposure to shifts in height and speed' (144). This interpretation was backed up by all the riders I watched and talked to, where gaps, ledges and rails became a conscious focus for the sensations they offered (see also Stevens 2006, 2007a, 2007b). When mobilised through risky and uncertain engagements, these inert features are transformed into desired places within the city. As Stevens (2007a) suggests, spaces such as edges are liminal thresholds that the body can negotiate in many different ways, a 'region of looseness and risk within a surrounding urban terrain which is flat and open' (82). Thus edges and gaps take on new functions through appropriation and repetition (Stevens 2007b: 147). For these riders, at least, such places are favoured not because of any visual or symbolic attachment; rather they are favoured because of the intense relations they offer when apprehended in the hybrid form of the bike-rider which serves to supplant certainty and banality with ephemerality and intensity in order to define affective and sensory meanings of place.

Perhaps even more surprisingly, for some riders like Phil – a 'flatland' BMX rider who regularly rides around the South Bank as well as near his home in Wood Green (north London) – the same intensity of feeling could be found in even the flattest and most featureless of spaces:

> When I meet Phil he's looking for a suitable place to ride. He opts for a deserted piece of tarmac very near to where the Trials riders were, in the shadow of the London Eye but flatter. There are a few free-runners here as well who have found a bit of street furniture, and a slow trickle of pedestrians. Phil straight away pulls a 'manual' and as he rides he adds other tricks into this; whipping the bike 360 degrees round the front fork to land back on it, riding it backwards from his front pegs and using his foot on the tyre to propel the bike.
>
> (field diary, 11 September 2005)

Similarly to the Trials[9] riders and in order to hone his skills, Phil evidently comes to the South Bank partly for its architecture, but in this case he chooses where to ride partly based on an absence of boundaries and obstacles, as he goes on to state when discussing balance as central to his love of BMX:

> I love that feeling of like balance and when you're peg-manualing and you can hold it for as long as there's road in front of you, then you've got the feeling of balance just carrying on and on and on forever.
>
> (Phil, September 2005)

Here Phil highlights the need for flat and featureless spaces which enable the duration of particular moves to be extended. Evidently, flat and deserted pieces of tarmac are not that hard to come by, as Phil showed me when riding from his home in Wood Green: his route to the South Bank itself took in detours of basketball courts and car parks as other examples of smooth concrete where he could practise without much of an audience. However, both Trials and BMX riders alike noted that a further reason to ride at the South Bank was to create a spectacle for the public to enjoy. Certainly the situation of the Shell Centre could not be better for this, as it is one of the key avenues through which visitors access the waterfront from Waterloo Station, which guarantees a constant stream of passers-by to view the Trials riding. The rationale for riding in public spaces was qualified by Trials riders Mark and David:

> Respect or appreciation from the public is quite a motivating part to riding; I am pretty well known as the rider in college.
>
> (Mark, 1 April 2006)

> (I ride) for a laugh and to put on a bit of a show when I know I'm being watched. Half the fun of Trials is that it's an unusual and relatively unknown sport, so people can be hostile if they think you're abusing a wall or intrigued and even amazed when they watch you ride.
>
> (David, 4 June 2006)

Ollie – a BMX rider – referring to riding outside the National Film Theatre, goes on to point to the fact that the reaction of the audience to something they've not seen is a central way through which riding becomes meaningful:

> I was riding down South Bank, end of last year, and there was a group of about 100 tourists and I was just like riding the banks and it was like they all started cheering and I was like what the fuck! It's the best thing that ever happened, I had a hundred people cheering me and I didn't even think I was doing anything.
>
> (Ollie, 5 July 2005)

Whether exaggerations or not, these accounts suggest passers-by do seem to be entertained by the performative and spectacular aspects of both Bike Trials and BMX, and on all the occasions that I observed the Trials riders, people would stop to watch, clap and take photographs and videos, as the following accounts illustrate:

> After some really good moves, I count about 15 people watching and photographing from above and about the same immediately around the riders. The crowd of onlookers seems genuinely impressed at the display going on and don't seem to mind that the occasional chip is being taken out of the street furniture.
>
> (field diary, 23 April 2006)

> As I approach the riders, two fifty-something ladies stop to watch as one rider flawlessly works his way round some stacks of paving slabs on his back wheel. Judging by the smiles on their faces I think they're nearly as impressed by the display as I am.
>
> (field diary, 11 September 2005)

Whilst Bike Trials, BMX and skateboarding are patently not intentionally included in the list of recreational and sporting activities and cultural pursuits that the SBEG seeks to promote, it is clear from their remit and the layout of the area that many of the South Bank's attractions are meant for visual apprehension, whether that be the landscape, the architecture, the commercial spaces, the street entertainers or, indeed, the Trials and BMX riders. Certainly the intention of the SBEG, in line with the 1999 White Paper,[10] is to make the area 'a desirable destination for cultural pursuits, business, education and leisure' with 'increased opportunities for culture, sport, recreation and shopping' and an emphasis on its promotion as a 'visitor destination' (SBEG 2009: 7). Similarly, the Commission for Architecture and the Built Environment (CABE 2000) response to redevelopment proposals implies an emphasis on the visuality and imagery of the area. It is clear from the CABE critique that the key tenets of this redevelopment are framed around pedestrian practices of socialising, consumption, display, and flaneury in order to make it a valued public space.

Despite this, there is no designated architecture or street furniture (beyond the undercroft, which I discuss later) at the South Bank for Trials and BMX riders, and the stance on skateboarding and similar activities, according to official documents, can be characterised – despite some acknowledgement of the skating community[11] – as negative. Yet the spaces of the South Bank can be conceptualised as what Stevens (2007b) describes as 'luxurious spaces'; that is, those spaces whose function is defined loosely, enabling them to serve a broad range of functions (199). Such spaces, they suggest, can take many forms, such as boundaries, ledges, walls, ramps or edges (8); their liminality and lack of definition affording unforeseen practices.

Drawing upon Rose's (2002) understanding of structures as performed, it is clear, as this account suggests, that the reactions of other users are vital in interpreting the actions of Trials and BMX riders as positive and appropriate for the area. Through stopping, watching, filming, photographing, clapping and cheering, passers-by perform different spatial meanings than those that might be inferred from the misuse of architectural features – in this case an acceptance of these forms of display – which serves to define them as appropriate. The positive reaction of passers-by forms part of the 'agencement' taking place because it serves to perform the rules and conventions of the place in a way that favours acceptance, and over-rides the formal (theoretical) exclusion of the riders. As Çalışkan and Callon suggest, the economic valuation of the social 'is a consequence of how competent and active people engage with specific things' (Çalışkan and Callon 2009: 388). In this case, it is the agencement between the riders, the physical space and the response of onlookers that imparts a value to Trials and BMX riding beyond the solipsistic and embodied value it gives to the riders themselves.

Affecting encounters

These riding styles are evidently not only engaging for the riders but also the spectators. However, what is important to understand in terms of value is how these riding practices become engaging and have the ability to affect passers-by to the extent that they will stop, watch, enjoy, applaud and film them. In order to do this, I briefly turn to literatures on affect. Affect has been variously defined in the literature, but the definition this chapter works with is that it is concerned with how emotions, sensations, atmospheres and feelings arise out of relational encounters between objects, spaces and people (Thrift 2004; Anderson 2006). As this suggests, the sources of affect – what moves us – are often from outside of the body (Thrift 2004: 60). This is important for understanding the ways in which the performances of Trials and BMX riders can affect others in the space.

One important element in shaping these encounters is the use and manipulation of environments in unexpected ways. As Ash has observed in relation to the design of computer games: 'designers are able to shape the potential of these affective encounters by altering users' game environment and by changing the rules that govern the users' engagement with that environment' (Ash 2010: 657). In fact, the game analogy is very fitting considering that Trials and BMX are in effect games played out on the city streets, one in which ledges become edges, and rails become points of balance.

Eleftheriotis (2012) argues that such unexpected and unintended use of space undermines the control of the viewer, by initiating 'unclear, contradictory, destabilizing or "catastrophic" spatial explorations' (82). As Eleftheriotis goes on to say, this destabilisation is the starting point for the opening up of these events along a continuum of certainty and uncertainty that has the power to affect because 'uncertainty [is] commonly perceived as the cause of intense anxiety' (ibid.). Massumi (2002) contends that it is in the gap between 'what is and what might be' that fosters contingent and imminent (uncertain) situations from which affects arise. This suggests that affects are emergent and imminent, rising out of uncertainty and the fact that many possibilities won't come to pass (Massumi 2002: 30–31). As Massumi goes on to note, there is an affective power in interruption that 'compresses into the movement underway potential movements that are in some way made present without being actualized' (Massumi 2002: 41). For example, once a Trials rider is balancing on an edge (or trying to hop on to an edge), what will happen next becomes uncertain, and this in turn can give rise to an anxiety (and excitement) in the viewer. Hence the capacity of trick cycling to affect resides in the 'probabilistic' in that it offers up multiple potential outcomes which it continually resolves. As Massumi states, 'viscerality is the perception of suspense' (2002: 61), and as a result the framework used to understand the affective power of such mobile practices is probabilistic rather than deterministic (Bissell 2010: 82).

Accordingly the 'dynamism' that such practices lend to the South Bank is a result of the atmospheres and affects they create through spontaneity, the unexpected, uncertain and probabilistic. Jones (2013) argues that in public spaces where access is (at least theoretically) restricted but only regulated according to behaviour, 'constraints on one's imaginative capacity – the ability to imagine, and project, oneself as "other", "*as-if*" – are loosened' (Jones 2013: 13). Indeed, it is the overflowing nature of play itself that is utilised in such instances: the imaginative and 'as if' qualities of these forms of probabilistic play enable others in the space to reimagine where they are as not just a place to shop or eat, but as a place of uncertain encounters and possibilities. It is precisely this sense of possibility (rather than instrumentally defined functioning) that gives dynamism to place. In the same way that the extravagant golf courses and pirate ships of the West Edmonton Mall give the idea that the mall is more than just a retail space (Goss 1993), the play of Trials, BMX riders and skaters defines the South Bank as a place beyond retail. This is possible because the probabilistic play of the Trials and BMX riders generates a dynamic behavioural surplus of 'what ifs'. Accordingly, the place itself is produced as one of uncertain and exciting encounter and meanings, overlaying extra-economic meanings onto the space. The value in this surplus is realised when people are entertained, dwell longer and consume more food, drink and 'stuff'. Evidently, as Jessop has argued, capital relations are articulations of both the economic and extra-economic which increasingly depend upon 'non-commodity forms of social relations' (Jessop 2000: 162).

Mobilising the social as agencement

Whilst one BMX rider succinctly summed up the feelings of pretty much all of the riders at the South Bank when he noted that, 'I can concentrate more when there's just nobody around,' it would be too easy to ignore the more mundanely social aspects of Bike Trials and BMX. According to Crouch (1998), academic accounts of the street have often overplayed its importance as a site of display, resistance, and the gaze, and downplayed its importance as a site of everyday geographical knowledge and leisure practice (160). Precisely because it forms such an everyday part of life, it is easy to overlook the fact that mundane social interaction is a key arena through which the meaning of particular activities is constructed and agencement occurs. Thus, whilst Bike Trials and BMX are all about the riding, they are also on a more mundane level all about socialising; at any one time there are likely to be as many riders chatting as there are actually riding. This interpretation is backed up by my own observations as well as by the riders themselves:

> I only ride on my own when I have to, like when I ride locally because there isn't anyone near me that rides. It's quite hard to ride alone, but it's fun when it's just to pass time or to chill and get some fresh air. There is more motivation to ride or try new things when you are around other riders, especially when there is someone of similar standards, since it provides good and healthy competition. [...] I still think there is a much closer bond between people that I ride with than others, considering we only meet every so often.
>
> (Mark, 1 April 2006)

When talking to riders it was clear – as Mark suggests – that riding with others meant they were more likely to try new things and challenge themselves, and were less likely to get bored because they could take a break to socialise. Whilst chat could range across many topics, much of it was focused on how to overcome particular challenges:

> I stop filming the rider in case I'm making him nervous and chat to David again, who is watching another rider attempt the same 4–5 ft vertical wall. He's not the only one, as two or three others are stood round lending advice to how the body should be positioned in order to achieve it. This is not idle advice, either, as one rider demonstrates with apparent ease by launching himself onto the ledge and then dropping back off.
>
> (field diary, 11 September 2005)

Moving beyond seeing these styles of riding solely as engagements between bike, body and architecture, the idea that riding is a social event is an important one that serves to flesh out the meanings of riding and the ways in which the dynamic and performing body is legitimated in such spaces. Just as Cuthbert and McKinnell (1997) describe Central Plaza in Hong Kong as being transformed into something human and vital by the weekend chatter of people, the spaces of the South Bank are recast as vibrant social spaces when filled with the chatter of riders.

The idea of the South Bank as 'a meeting place' is explicitly recognised and commodified in a recent South Bank Marketing Group (SBMG) campaign Lambeth Council report (2009: 6). The South Bank has thus been remodeled as a profoundly social space into which the 'social event' of riding Trials and BMX sits quite comfortably. I would suggest that acknowledging this is important in understanding why riders are often tolerated, if not openly welcomed, on the South Bank. Moreover, it is clear from talking to riders that they too are attracted to the vibrance of the area as Phil attests:

> I like places more because of what they offer in terms of actual riding. Having said that, I do come to the South Bank for the aesthetics, I suppose I do, I mean the atmosphere is nicer with the music and all the people and the graffiti. They often have a band on and it's wicked because it breaks up your riding. There's this guy who comes down called Sebastian and he comes down with his guitar and reverb box and sometimes I'll ride just behind where he plays because he is the bollocks, man, and he's so into it and it's inspirational to hear him play, and there's another guy down there who I was just listening to. He plays the Jemba, some West African guy and he plays it really well and sometimes he's down here with a load of South Africans and it's wicked man. I'll just chill down with them for a little bit and it's inspirational. I might ride at Finsbury Park in the basketball courts for two hours and it gets boring. Down here you can come and have a break and look what you've got – it's amazing.
>
> (Phil, September 2005)

Phil's account of wanting to soak up the sights and sounds of the South Bank illustrates that rather than setting himself apart (or being set apart) from other citizens, he and other riders are just as susceptible to the lure and buzz of the area as they are responsible for creating the buzz of the area. In this sense, the riders are just as much consumers of its aesthetic as anyone else. Much like the spectacle of riding, the social side of riding performs the intended uses of the South Bank, actively producing it as a social space. Far from being resistant to the dominant meanings of the area, I would argue that the social conduct of Bike Trials and BMX at the South Bank reflect the desire of riders to be part of the 'mainstream' through taking part in everyday urban life (cf. Donnelly (2008) and Rinehart (2008) in relation to skateboarding). Whilst the riders' relative lack of purchasing in the area (they are generally young and have less money than the target market) might position them as undesirable consumers, they are accepted because they add to the production of the area in other ways. In line with the previous section, the performance of the social aspects of Trials and BMX riding represents another aspect of agencement, because it is a central way through which qualities of a practice deemed useful and congruent to the aims of the area are foregrounded. At the same time, and as I demonstrate in the next section, riding is managed out of spaces where it does not reproduce the desired qualities of place.

Managing and policing as forms of agencement

It is clear from the previous sections that particular aspects of riding (its aesthetic dynamism and vocal chatter) are foregrounded and accepted in areas of the South Bank precisely because they complement the aims of the area. Thus far, this represents a kind of 'management by social norm', where practices that could be deemed transgressive are accepted as long as pro-social qualities are foregrounded and anti-social qualities are backgrounded. This more dispersed kind of social agencement is also complemented by more overt management strategies to ensure that only desirable variants of riding are materialised, and only in the 'right' places.

As could be expected, skateboarding and riding in London are more or less accepted in particular spaces. Previous research on skateboarding has shown, for example, that there is a clear distinction in the levels of social control experienced in public spaces and purpose-built skate parks (Chiu 2009: 26). Accordingly, whether a rider is deemed to be in place or out of place is not absolute but depends on a range of factors, including where they are, what they are trying to achieve, what else they might be doing, their numbers and who is policing them. Balancing these factors results in a constant movement between more or less sanctioned spaces (Vivoni 2009: 131).

It became clear from talking to riders that where they rode was initially dictated by what they were trying to achieve: to get some uninterrupted practice, or to be seen and put on a display for the public, as Mark and Phil illustrated when explaining their choices:

> Attention is good sometimes, but it normally puts off less able riders. I personally don't really like it when people aren't riding because of a crowd, so I don't like staying in public for too long, although it doesn't bother me that much. (If we ride) in less public areas, it puts less pressure on yourself and the other riders, I guess, making it more of a learning session than a free show or competition for attention.
>
> (Mark, April 2006)

> Yeah they're practising in public which … it's not that I'm bothered but I can concentrate more when there's just nobody around. When I used to ride in Birmingham for those years, there's a place called the ICC (International Convention Centre) and I used to go to the car park there 'cos it was always empty when they first built it and I just used to ride in the underground car park … every night. I just used to go there and ride, put my headphones on and just ride my bike.
>
> (Phil, September 2006)

Evidently, the choice to ride on the South Bank suggests that riders have made a decision to be on display, rather than solely practising their skills. There are of course other places where riders could be on display and where the street furniture created great challenges for them. However, their accounts of these places suggested that they were effectively policed out of them:

> I'm in Trafalgar Square waiting for Ollie to arrive. Ten minutes later, Ollie scoots into view on his BMX. The first thing he says is that he

> thinks he's about to be thrown out. Before I've even had a chance to ask why, the security guard walks over, looking none too pleased. I can't recall the conversation exactly but it's obvious that this is not the first run-in that the 'heritage warden' has had with Ollie. He angrily asks him why he rides his bike down the stairs – is it just to wind him up? After the security guard leaves, I ask Ollie why he was so angry. He says that he and his friends do make the bloke's life hard – they often come down here after a few drinks riding the stair sets.
>
> (field diary, 21 July 2005)

> David checked for security in the car park initially before they all came over to give the all-clear. Michael says that they don't have much trouble around here (the Shell Centre) but that the City and St Paul's have become no-go areas for them due to security and policing.
>
> (field diary, 11 September 2005)

As these accounts suggest, places like Trafalgar Square and St Paul's are very much policed as pedestrian spaces, where acceptable activities such as sight-seeing and sitting should not be interrupted by spectacles such as BMX and Trials riding. If riders ignore these social and design codes around intended use, the security enforce them unequivocally and manage these inappropriate uses out of the spaces.

However, it is also clear that certain places are managed much less strictly, suggesting a tacit acceptance that spectacle is acceptable and even desirable in some spaces. On the South Bank, the only space that could be deemed 'officially' sanctioned and largely unpoliced is that of the National Film Theatre undercroft, which I discuss later. In all other locations apart from this, riders encountered a varied enforcement of rules, with some public areas enforced more stringently. This was made clear to me each time I accompanied the Trials riders. They were asked on a number of occasions to move on by private security at the Shell Centre and by residents and passers-by in other parts of the South Bank:

> Yeah security do give us some flak, but I'll give them flak back in return. From experience they won't call the police, and if they do you've got ten minutes before they arrive and ask you politely to go away ... whilst they're only following orders, if I know I'm not doing any damage/harm, only the police are going to tell me to stop.
>
> (David, 4 June 2006)

> They [security] always tell us to move on because we normally ride on private property, but some are much more appreciative and polite than others, sometimes I would get compliments which normally makes me feel more obliged to move on. Most are friendly enough, though. Most are quite taken aback by our bikes not having seats, including the police!
>
> (Andrew, 1 April 2006)

As these accounts suggest, the attitude of the riders to authority varied depending on who was asking them to move, and particularly in David's comments, there is certainly evidence that authority is being actively resisted. However, the point that I want to emphasise here is that the deterrent of private security is particularly weak. On the occasions I was with riders, private security did not call the police or come back to check whether their instructions had been obeyed. Coupled with a lack of enforcement powers, private security do little to perform the South Bank as out of bounds. Such weak management contrasts with that found by the riders in the city and Trafalgar Square, where appropriate uses were strongly enforced. One implication of this is that some transgressive practices are tolerated, particularly if those transgressions help to produce the desired buzz and spectacle of the area.

Building upon a typology of space developed by Flusty (2000), Stratford (2002) distinguishes between different levels of control in public spaces, but ultimately positions all public spaces (except those of the skatepark) as exclusionary towards skateboarders and similar urban practices such as BMX. Similarly, Borden argues that similar levels of enforcement are encountered in all public skate spots, where skateboarders will be treated 'like the homeless' because they are not the kinds of consumers that contribute to the economy of the area (Borden 2001: 253). According to Grafham (2002), the result of such binary thinking is that skaters will be displaced from 'adopted' public spaces and located to purpose-built skate parks.

However, the evidence presented here suggests a somewhat different strategy, with police and security helping to contain riding in areas (such as the South Bank) where qualities of spectacle and sociality help to produce the desired buzz of the area, whilst strictly maintaining the pedestrian and more sedate qualities of other areas (such as St Paul's and The City) through zero-tolerance management strategies making life harder in other areas such as St Paul's. Whilst the logic of this policing may also be linked to the materialisation of counter-terrorism strategies at the borough level than preserving public property (see Coaffee 2009), it is just as likely to represent a way of maintaining particular place images, as Belina and Helms (2003) have argued. Whichever, it is clear that there are different levels of enforcement being practised in different locations and it seems unlikely that the much weaker enforcement on the South Bank is due solely to reduced terror threat.

Bringing this discussion back to the key argument of the chapter, I argue that the management of riders on the South Bank represents another way in which agencement occurs through the definition of roles and qualities. Through weak enforcement, riders come to understand that even if they do not spend money in the area, as long as they don't act anti-socially (e.g. take drugs, fight, etc.), their architectural transgressions will be tolerated. As such, management acts as a kind of filter, where certain qualities are allowed to flourish whilst others are not. It is no accident that the qualities that are allowed to flourish are those that facilitate the production of the South Bank as a vibrant and spontaneous place. Thrift has argued that in the drive for cities to have a buzz and become creative and inventive, affective and aesthetic practices are being manufactured because they are an economic commodity (Thrift 2004: 58). The example here of riding on the South Bank and the way it is accommodated because of its ability to produce desirable atmospheres seems to perfectly embody this argument.

Material agencements

The final agencements I want to discuss relate to the spaces that are designated for such spectacular practices. The examples I have covered so far all relate to the 'found' spaces and the uncertain management and negotiation of the rules and conventions regarding appropriate conduct in them. However, it could be argued that it would be more desirable for Trials and BMX riders to have a permanent home at the South Bank because, as one architectural critic noted of initial attempts to discourage skateboarding and riding, 'the South Bank Centre's biggest mistake was to discourage the skateboarders and mountain bikers. It was created as a vibrant social space – to keep that, you need to have some activity there' (Wilcox 2002).

Such a position is reflected in more recent developments of the area that occurred after the fieldwork for this chapter was completed. In 2013, the only 'official' skating and riding space on the South Bank (the undercroft of the NFT) was under threat due to redevelopment plans for the new Festival Wing extension, which proposed the complete removal of the undercroft skate area and its relocation further along the South Bank. Under these original proposals, the undercroft would be turned into retail units (Escobales 2013) and the skaters relocated to Hungerford Bridge, about 100 metres from the original site. In line with other current local government strategies, the focus here is on displacing undesirable practices from central public spaces to peripheral semi-private spaces with little acknowledgement of the importance of display to the conduct of such practices, as Grafham (2002) notes:

> The subculture that we are focusing on largely has its roots in the streets and urban architecture. In providing a permanent skate park, or purpose-built facilities, we are offering an alternative to the 'adopted' sites in the city center. It is a positive measure by the council with which to support anti-skate measures elsewhere in the city.
>
> (in Rogers 2006: 113)

The strategy of reducing 'skaters on the streets' through displacement and provision (Rogers 2006: 113) is a familiar one premised in the first instance upon an understanding of skateboarding as an encounter between the body and architecture. Whilst this is of course the case, the problem with such a strategy is its basis in a narrow conceptualisation of skateboarding – and analogous practices such as Trials and BMX – as only an embodied encounter with architecture. This is compounded by the classification of such practices – as Grafham does – as subcultural. Such a reading suggests – in a classic Centre for Contemporary Cultural Studies (CCCS) vein – that riders want to set themselves apart and resist commodified mainstream culture (Weinzierl and Muggleton 2003: 8), a reading that the more dialectical account presented here opposes.

Rose suggests that, by framing practice in terms of resistance, we risk reinforcing 'the very systems of power that practices of resistance are thought to undermine' (Rose 2002: 388). A clear example of such a misdiagnosis of resistance in action is the proposed strategy of South Bank's managers to replicate urban architecture in officially sanctioned spaces in the belief that this will provide riders with everything that adopted public spaces do. On the contrary, like many other citizens, riders are attracted to places like the South Bank because they want to socialise, see and be seen, and be part of 'mainstream' culture. As a result, strategies that seek to remove riders and skateboarders from these areas to peripheral areas are likely to meet with limited success because these (officially sanctioned) areas do not address all their needs. Skateparks and purpose-built facilities neglect the fact that much of the point of skateboarding, BMX and Bike Trials is also about display, spectacle, sociability and buzz – all things that are experienced very differently at the skatepark.

After a 17-month campaign by 'Long Live South Bank', and with the surprise backing of then London Mayor Boris Johnson, in 2014 the initial plans to move the riders/skaters were adapted to keep the undercroft and allow them to stay (Brown 2014). In the adapted plans the Southbank Centre agreed to keep the undercroft, 'open for use without charge for skateboarding, BMX riding, street writing and other urban activities' (Brown 2014). Following this announcement, in 2018, City Hall granted the undercroft a £700,000 grant that would expand by one-third the area they currently use, as well as improve lighting, reinstate original features and provide a creative education centre (De Peyer 2018). The South Bank Centre CEO stated that it would give skaters and BMXers access to 'newly opened up, and restored, sections of the undercroft' (Bedell, quoted in de Peyer 2018).

What is interesting is that the arguments promoted by the groups charged with revitalising the South Bank revolve around the links between skating and BMX culture and their contribution to urban vitality, with many headlines and commentaries proclaiming that 'grassroots culture [is] still worth fighting for'. Indeed, the reversal in fortunes and the dominant narrative suggest a realisation that what the skateboarders, BMX and Trials riders bring to the area is very much a part of the official imagining of it as a culturally diverse and dynamic area. Mould (2013), however, notes in his commentary on the redevelopment that:

> the fact that the South Bank initially tried to stop them [skaters and BMX riders] is outweighed by their commendable decision to let them

> continue (when this happened though is unclear). One suspects though that once 'they' realised that spectators flocked to the area to see the skaters and BMX riders in action and then emptied their pockets in the surrounding bars and amenities, they were more than happy to let them continue.
>
> (Mould 2013)

Mould goes on to note that in the redevelopment plans the area around the undercroft is to be 'lined with retail outlets' (ibid.). As a result, Mould is scathing of the redevelopment's cultural value and sincerity, arguing instead that these spectactular forms of movement are being spatially accommodated solely because of their role in 'entertailing'.

The point I want to draw attention to is that this accommodation represents a material and spatial agencement that aligns cultural performance with economic logic. Çalışkan and Callon (2009) emphasise the importance of institutions in the (economic) formatting of practice, stating that they 'enhance and bring into existence the competencies which can potentially exist in human beings; institutions afford these competencies the possibility of being enacted' (Çalışkan and Callon 2009: 380). Certainly, by curating a space and cultural narrative (e.g. skating/BMX as 'urban treasure'), the institutions that govern the South Bank enable alternative playful mobile practices to gain currency, and literally gain value. Activities excluded from this narrative remain designated and classified as unwelcome activity. However, once included in it, activities become a resource and another curated 'event' that value can be extracted from as adding dynamism to the area.

Such accommodation of previously disregarded practices prompts the question of whether such 'deviant' practices would be so welcome if they did not produce a behavioural surplus that can be readily harnessed to serve the broader economic goals of the area. Certainly, it appears that no one is falling over themselves to accommodate other equally transgressive but less carnivalesque practices in the area (such as begging or rough sleeping), suggesting that whilst there is a cultural benefit to keeping BMX and skating at the South Bank, the over-riding and unspoken imperative is one that recognises its contribution to 'entertailing' and seeks to harness this. As Boland has argued, whilst 'the presence of gangs of youths, prostitutes, drug addicts, alcoholics, and homeless people do not conform to a city's sexy new image' (Boland 2013: 261) and thus remain excluded, 'entrepreneurial rule' accommodates more acceptable activities through a series of managerial, social and spatial agencements.

Conclusions

Through ethnographic engagement with the everyday practices of BMX and Bike Trials at London's South Bank, in this chapter I have attempted to highlight the ways in which particular forms of mobile conduct are legitimated in order to generate and realise a behavioural surplus. Through a series of managerial, social and spatial agencements I have shown how the social practice of Trials and BMX riding is aligned with the aims of the South Bank redevelopment so that the dynamism, uncertainty and spontaneity that characterise the riders' movements can add value to the retail environment. In doing so, I have tried to focus attention on the conduct and practice of riding in public spaces rather than on the body/architecture dialectic (Borden 2001; Flusty 2000) or the subcultural elements of practice (Beal 1995; Vivoni 2009). As a result this chapter has illustrated a number of reasons why spatial practices previously theorised as resistant might more accurately be reframed as 'conforming' to the social, cultural and economic, if not architectural, rules and conventions of particular public spaces.

Donnelly (2008) notes how the skateboarding media works hard to create the illusion of resistant and alternative identities, but argues that, in reality, these are highly commodified practices with much in common with mainstream culture (see also Rinehart 2008). Similarly, previous academic accounts, such as those of Beal (1995), Borden (2001) and Flusty (2000), have served to position unsanctioned spatial practices as resistant. Here, however, I have argued that whilst the body–architecture dialectic may be inappropriate, the spectacular and social nature of BMX and Bike Trials can be congruent with the ideals enshrined in retail and culture-led redevelopment plans if filtered and harnessed accordingly. Wider practices of spectating, socialising, management, narrative and place perform the South Bank as largely conducive to these styles of riding, illustrating the inability of structures to endure separate from performance (Rose 2002).

More broadly, I argue that the fate of Trials and BMX at the South Bank helps explain how particular geographies of urban cycling unfold in towns and cities. In particular, why we see particular manifestations of urban cycling and not others, and why we see these occurring in particular places and not others. In this instance, whilst such playful forms of cycling are written out of instrumental road spaces as inefficient and dangerous, they may be facilitated and tolerated in retail-led public spaces because it is recognised that their spectacular nature can form part of the entertailing that makes the area distinct and desirable to consumers. However, this has a distinct geography, with these styles of riding only being tolerated in public spaces where – as Thrift (2004) has argued – such affective practices contribute to benefitting the economy of the area through the production of buzz, dynamism and spontaneity. Once again we see the nature and place of contemporary manifestations of urban cycling being shaped by narrow economic goals.

Notes

1 This chapter is based on a published paper: https://journals.sagepub.com/doi/abs/10.1068/a43149

2 Bike Trials was originated in Spain and the US by two well-known Motocross and Trials riders, Pedro Pi (national Motocross champion of Spain) and Eddy Kessler, respectively. In common with its motorised counterpart, the goal of Bike Trials is for the rider to traverse a series of obstacles without the feet touching the ground. Whilst motorised Trials riding has traditionally taken place in 'natural' environments, the advent of non-motorised Bike Trials has allowed the practice to flourish in urban environments.

3 BMX, or Bicycle Moto-Cross, originated in Southern California in the late 1960s and became popular throughout the 1970s, finally 'arriving' in the UK in the late 1970s and early 1980s, where it was very popular for a short space of time (Barraclough and Buckden 1982: 12). It continues to be popular as a 'subcultural' form of riding and has, if anything, seen a gradual increase in its popularity over the past ten years.

4 A significant number of authors have highlighted the transgressive and resistant nature of skateboarding as a practice. In one of the earliest sociological accounts of skating, Beal (1995) positions it as a culture of social resistance that challenges dominant norms and design codes. Stratford (2002) suggests that skating goes beyond what is considered rational and legitimate use of space (Stratford 2002: 192), and according to Vivoni (2009), 'skateboard traces most often signal an unsanctioned engagement with the built environment that undermines prescribed uses, social norms, and legal mandates founded on the sanctity of property value and intent on regulating social relations in public space' (Vivoni 2009: 141). As a result skate-spots are 'set against the material expressions of contemporary trends in spatial regulation and urban governance' (ibid.). With a focus on the architectural and subcultural, such accounts maintain that skateboarding is inherently and consistently oppositional.

5 In the case of the South Bank, 'the firm' is not a single entity but rather a collectif of stakeholders.

6 The site which is owned by Shell was closed in 2007 with a view to redevelopment pending the findings of a planning enquiry (Farrer Huxley Associates 2004: 136).

7 Whilst all cycling is officially outlawed on the South Bank, commuting-style cycling may be less targeted by security patrols due to the fact that it does not involve congregations of 'youth' (see SBEG 2009).

8 The SBEG is one of a number of community groups that has emerged in the last decade in order to help deliver the Waterloo Development Framework. The SBEG represents businesses and major organisations and is part of a hierarchical structure of groups that includes actors at the borough, regional and national level. Notably, the SBEG also operates subgroups such as the South Bank Marketing Group.

9 On the occasions that I accompanied the riders, their route followed a relatively set path from the Shell Centre out towards the area around St Paul's Cathedral via the Tate Modern. It is, of course, a route that is continuously evolving over time as new obstacles are found close to the 'usual' route. The accounts I represent here all focus on the area around London's South Bank.

10 Rogers' 1999 *Urban White Paper* positions public space as 'somewhere to relax and enjoy the urban experience, a venue for a range of different activities, from outdoor eating to street entertainment; from sport and play to a venue for civic or political functions; and most importantly of all as a place for walking and sitting out' (Rogers 1999).

11 However, the *Space for All* report states that 'the community of skateboarders, although considered a nuisance by some parties, would benefit from having a permanent site' (Farrer Huxley Associates 2004: 39).

References

Anderson, B. (2006) Becoming and being hopeful: Towards a theory of affect, *Environment and Planning D: Society and Space*, 24(5), pp. 733–752.

Arnold, M. and Reynolds, K. (2003) Hedonic shopping motivations, *Journal of Retailing*, 79, pp. 77–95.

Ash, J. (2010) Architectures of affect: Anticipating and manipulating the event in processes of videogame design and testing, *Environment and Planning D: Society and Space*, 28, pp. 653–671.

Barraclough, G. and Buckden, I. (1982) *Riders Ready … Pedals Ready … Go: BMX in Britain* (Humberside: Reflex Publications).

Beal, B. (1995) Disqualifying the social: An exploration of social resistance through the subculture of skateboarding, *Sociology of Sport Journal*, 12, pp. 252–267.

Belina, B. and Helms, G. (2003) Zero tolerance for the industrial past and other threats: Policing and urban entrepreneurialism in Britain and Germany, *Urban Studies*, 40(9), pp. 1845–1867.

Biddulph, M. (2011) Urban design, regeneration and the entrepreneurial city, *Progress in Planning*, 76, pp. 63–103.

Bissell, D. (2008) Comfortable bodies, sedentary affects, *Environment and Planning A*, 40(7), pp. 1667–1712.

Bissell, D. (2010) 'Placing affective relations: Uncertain geographies of pain', in Anderson, B. and Harrison, P. (eds) *Taking Place: Non-representational theories and geography* (Farnham: Ashgate), pp. 79–98.

Blume, M. (2001) A visionary architect with designs on the past, *International Herald Tribune*, 28 July. Available at: http://iht.com/articles/2001/07/28/mather_ed2_.php [accessed April 2005].

Boland, P. (2013) Sexing up the city in the international beauty contest: The performative nature of spatial planning and the fictive spectacle of place branding, *Town Planning Review*, 84(2), pp. 251–274.

Borden, I. (2001) *Skateboarding, Space and the City: Architecture and the body* (Oxford: Berg).

Brindley, T., Rydin, Y. and Stoker, G. (1996) 'Popular planning: Coin Street London', in Campbell, S. and Fainstein, S. (eds) *Readings in Planning Theory* (Malden, MA: Blackwell), pp. 296–317.

Brown, M. (2014) Skaters will stay at the undercroft as London's Southbank Centre gives in, *The Guardian*, 18 September. Available at: www.theguardian.com/uk-news/2014/sep/18/skaters-southbank-centre-undercroft-london-remains-open [accessed September 2019].

Çalışkan, K. and Callon, M. (2009) Economization, part 1: Shifting attention from the economy towards processes of economization, *Economy and Society*, 38(3), pp. 369–398.

Çalışkan, K. and Callon, M. (2010) Economization, part 2: A research programme for the study of markets, *Economy and Society*, 39(1), pp. 1–32.

Callon, M. (2007) An essay on the growing contribution of economic markets to the proliferation of the social, *Theory, Culture and Society*, 24(7–8), pp. 139–163.

Callon, M., Méadel, C. and Rabeharisoa, V. (2002) The economy of qualities, *Economy and Society*, 31(2), pp. 194–217.

Chain Store Age (2003) Waterfront retailing: Branson Landing brings retail and entertainment to Lake Taneycomo, *Chain Store Age*, 79(12), p. 160.

Chiu, C. (2009) Contestation and conformity: Street and park skateboarding in New York City public space, *Space and Culture*, 12(1), pp. 25–42.

Cloke, P., May, J. and Johnsen, S. (2010) *Swept up Lives? Re-envisioning the homeless city* (Chichester: Wiley-Blackwell).

Coaffee, J. (2009) *Terrorism, Risk and the Global City: Towards urban resilience* (Aldershot: Ashgate).

Commission for Architecture and the Built Environment (CABE) (2000) *South Bank Masterplan – CABE Design Review*. Available at: www.cabe.org.uk/default.aspx?contentitemid=888&field=filter&term=London&type=7 [accessed April 2006].

Cosgrove, D. (1996) Windows on the city, *Urban Studies*, 33(8), pp. 1495–1498.

Cresswell, T. (1993) Mobility as resistance: A geographical reading of Kerouac's 'On the Road', *Transactions of the Institute of British Geographers*, 18, pp. 249–262.

Cresswell, T. (1996) *In Place, Out of Place: Geography, ideology and transgression* (Minneapolis: University of Minnesota Press).

Cresswell, T. (2006) *On the Move* (Oxford: Routledge).

Crouch, D. (1998) 'The street in the making of popular geographical knowledge', in Fyfe, N. (ed.) *Images of the Street: Planning, identity and control in public space* (London: Routledge), pp. 160–175.

Cuthbert, A. and McKinnell, K. (1997) Ambiguous space, ambiguous rights: Corporate power and social control in Hong Kong, *Cities*, 14(5), pp. 295–311.

Darley, A. (2000) *Visual Digital Culture: Surface play and spectacle in new media genres* (London: Routledge).

Davis, M. (1992) *City of Quartz* (London: Verso).

Davis, M. (1998) *Ecology of Fear: Los Angeles and the imagination of disaster* (New York: Metropolitan Books).

De Peyer, R. (2018) South Bank's famous undercroft skatepark to be given huge extension, *Evening Standard*, 11 March. Available at: www.standard.co.uk/news/london/south-banks-famous-undercroft-skate-park-to-be-given-major-extension-a3785671.html [accessed August 2019].

Dodge, P. (1996) *The Bicycle* (Paris/New York: Flammarion).

Donnelly, M.K. (2008) 'Alternative and mainstream: Revisiting the sociological analysis of skateboarding', in Atkinson, M. and Young, K. (eds) *Tribal Play: Subcultural journeys through sport* (Bingley, W. Yorks: JAI Press), pp. 197–214.

Dovey, K. (2005) *Fluid City: Transforming Melbourne's urban waterfront* (Sydney: University of New South Wales Press).

Dovey, K. and Stevens, Q. (2005) 'Appropriations", in Dovey, K. (ed.) *Fluid City: Transforming Melbourne's urban waterfront* (Sydney: University of New South Wales Press), pp. 67–80.

Eleftheriotis, D. (2012) *Cinematic Journeys: Film and movement* (Edinburgh: Edinburgh University Press).

Escobales, R. (2013) Skateboarding's South Bank home to be turned into retail units, *The Guardian*, 12 April. Available at: www.theguardian.com/uk/2013/apr/12/skateboarding-south-bank [accessed September 2019].

Farrer Huxley Associates (2004) *Space for All: Waterloo open space audit* (Bankside Open Space Trust).

Flusty, S. (2000) Thrashing downtown: Play as resistance to the spatial and representational regulation of Los Angeles, *Cities*, 17(2), pp. 149–158.

Foucault, M. (1995) *Discipline and Punish: The birth of the prison*, trans. Alan Sheridan (New York: Random House).

Foucault, M. (2010) *The Birth of Bio-politics: Lectures at the college de France 1978–79* (Basingstoke, UK: Palgrave Macmillan).

Franck, K. and Stevens, Q. (eds) (2007) *Loose Space: Possibility and diversity in urban life* (Oxford: Routledge).

Fyfe, N. (ed.) (1998) *Images of the Street: Planning, identity and control in public space* (London: Routledge).

Goss, J. (1993) The 'magic of the mall': An analysis of form, function, and meaning in the contemporary retail built environment, *Annals of the Association of American Geographers*, 83(1), pp. 18–47.

Grafham, K. (2002) *Evaluation of City Centre Skate Spots: Report and recommendations for Newcastle City Council* (Newcastle upon Tyne: Tribe Enterprises Ltd).

Harvey, D. (1989) From managerialism to entrepreneurialism: The transformation in urban governance in late capitalism, *Geografiska Annaler*, 71B(1), pp. 3–17.

Harvey, D. (2005) *A Brief History of Neoliberalism* (Oxford: Oxford University Press).

Hill, J. (ed.) (1998) *Occupying Architecture: Between the architect and the user* (London: Routledge).

Hill, J. (2003) *Actions of Architecture: Architects and creative users* (London: Routledge).

Imrie, R. and Street, E. (2009) 'Regulating design': The practices of architecture, governance, and control, *Urban Studies*, 46, pp. 2507–2518.

Jessop, B. (2000) 'The state and the contradictions of the knowledge-driven economy', in Bryson, J., Daniels, P., Henry, N. and Pollard, J. (eds) *Knowledge, Space, Economy* (London: Routledge), pp. 63–78.

Jessop, B. (2002) *The Future of the Capitalist State* (Oxford: Polity Press / Malden, MA: Blackwell).

Jones, A. (2013) A tripartite conceptualisation of urban public space as a site for play: Evidence from South Bank, London, *Urban Geography*, 34(8), pp. 1144–1170.

Jones, A. (2016) *Orchestrated Public Space: The curatorial dimensions of the transformation of London's Southbank Centre* (LSE Research Online). Available at: http://eprints.lse.ac.uk/68373/1/Jones_Orchestrated%20public%20space.pdf [accessed August 2019].

Kearns, G. and Philo, C. (eds) (1993) *Selling Places* (Oxford: Pergamon).

Kern, K. (2008) 'Heterotopia of the theme park street', in Dehaene, M. and De Cauter, L. (eds) *Heterotopia and the City: Public space in a postcivil society* (London: Routledge), pp. 105–115.

Krivy, M. (2018) Towards a critique of cybernetic urbanism: The smart city and the society of control, *Planning Theory*, 17(1), pp. 8–30.

Lambeth Council (2009) *Waterloo Area Supplementary Planning Document*. Available at: www.lambeth.gov.uk/NR/rdonlyres/A7287D47-58FD-4874-ADB8-1CF2CF1FCF4E/0/WaterlooSupplementaryPlanningDocumentJune2009.pdf

Lees, L. (1998) 'Urban renaissance and the street: Spaces of control and contestation', in Fyfe, N. (ed.) *Images of the Street: Planning, identity and control in public space* (London: Routledge), pp. 236–253.

Lees, L. (2004) *The Emancipatory City: Paradoxes and possibilities* (London: Sage).

Lorimer, H. (2005) Cultural geography: The busyness of being more-than-representational, *Progress in Human Geography*, 29(1), pp. 83–94.

Lorimer, S. and Marshall, S. (2015) Beyond walking and cycling: Scoping small-wheel modes, *Proceedings of the Institution of Civil Engineers – Engineering Sustainability*, 169(2), pp. 58–66.

Lowe, M. (2005) The regional shopping centre in the inner city: A study of retail-led urban regeneration, *Urban Studies*, 42(3), pp. 449–470.

MacLeod, G. and Ward, K. (2002) Spaces of Utopia and dystopia: Landscaping the contemporary city, *Geografisker Annaler*, 84b(3–4), pp. 153–170.

Markham, S. (1996) 'Nick Kaufmann: On a wheel against time', in Van der Plas, R. (ed.) *Proceedings of the 7th International Cycle History Conference* (Anaheim, CA: KNI), pp. 65–73.

Marshall, R. (ed.) (2001) *Waterfronts in Post-industrial Cities* (London: Spon Press).

Massumi, B. (2002) *Parables for the Virtual: Movement, affect, sensation* (Durham, NC/London: Duke University Press).

Matusitz, J. (2012) Cathedrals of consumption, *Wiley-Blackwell Encyclopedia of Globalization*, https://doi.org/10.1002/9780470670590.wbeog064.

May, J. (2000) Of nomads and vagrants: Single homelessness and narratives of home as place, *Environment and Planning D: Society & Space*, 18, pp. 737–759.

Middleton, J. (2009) 'Stepping in time': walking, time, and space in the city, *Environment and Planning D: Society and Space*, 41, pp. 1943–1961.

Miller, J.C. (2013) The spatial dialectics of modernity and retail affect at Abasto Shopping, Buenos Aires, Argentina, *Urban Geography*, 34(6), pp. 843–863.

Mould, O. (2013) *Save the South Bank's Undercroft*. Available at: https://tacity.co.uk/2013/04/24/save-the-south-banks-undercroft/ [accessed August 2019].

Pinder, D. (2002) In defense of Utopian urbanism: Imagining cities after the end of Utopia, *Geografiska Annaler*, 84b(3–4), pp. 229–241.

Rendell, J. (1998) 'Doing it, (un)doing it, (over)doing it yourself: Rhetorics of architectural abuse', in Hill, J. (ed.) *Occupying Architecture: Between the architect and the user* (London: Routledge), pp. 230–246.

Rinehart, R. (2008) 'ESPN's X Games: Contests of opposition, resistance and conformity, and negotiation", in Atkinson, M. and Young, K. (eds) *Tribal Play: Subcultural journeys through sport* (Bingley, UK: JAI Press), pp. 175–195.

Ritzer, G. (1999) *Enchanting a Disenchanted World: Continuity and change in the cathedrals of consumption* (New York: Sage).

Robins, K. (1991) Prisoners of the city: Whatever could a postmodern city be?, *New Formations*, 15, pp. 1–22.

Rogers, R. (1999) *Towards an Urban Renaissance: Final Report of the Urban Task Force Chaired by Lord Rogers of Riverside* (London: Department of the Environment, Transport and the Regions).

Rogers, P. (2006) Young people's participation in the renaissance of public space: A case study in Newcastle upon Tyne, UK, *Children, Youth and Environments*, 16(2), pp. 105–126.

Rose, M. (2002) The seductions of resistance: Power, politics, and a performative style of systems, *Environment and Planning D: Society and Space*, 20, pp. 383–400.

SBEG (2009) *South Bank Employer's Group Business Plan 2009–2012*. Available at: www.sbeg.co.uk/documents/SBEGBusPlan09updatedJULY09.pdf [accessed June 2010].

Smith, N. (2002) New globalism, new urbanism: Gentrification as global strategy, *Antipode*, 34(3), pp. 427–450.

South Bank BID (2019) *Renewal Business Plan*. Available at: https://southbankbid.co.uk/wp-content/uploads/2019/05/South-Bank-Business-Plan-Final.pdf [accessed November 2019].

South Bank Centre (2007) *Transforming South Bank Centre*. Available at: www.southbankcentre.co.uk/about-us/transforming-southbank-centre [accessed July 2020].

South Bank Employers Group (2009) *South Bank Employer's Group Business Plan 2009–2012*. Available at: www.sbeg.co.uk/documents/SBEGBusPlan09updated JULY09.pdf.

Spinney, J. (2006) A place of sense: A kinaesthetic ethnography of cyclists on Mt Ventoux, *Environment and Planning D: Society and Space*, 24(5), pp. 709–732.

Spinney, J. (2007) 'Cycling the city: Non-place and the sensory construction of meaning in a mobile practice', in Horton, D., Rosen, P. and Cox, P. (eds) *Cycling and Society* (Aldershot, UK: Ashgate), pp. 25–46.

Spinney, J. (2008) *Cycling the City: Movement, meaning and practice*, unpublished PhD thesis (Royal Holloway University of London).

Spinney, J. (2010) 'Mobilising sustainability: Partnership working between a pro-cycling NGO and local government in London (UK)', in Fudge, S., Peters, M. and Jackson, T. (eds) *Low Carbon Communities: Imaginative approaches to combating climate change* (Cheltenham, UK: Edward Elgar), pp. 89–107.

Staeheli, L. and Mitchell, D. (2006) USA's destiny? Regulating space and creating community in American shopping malls, *Urban Studies*, 43, pp. 977–992.

Stevens, Q. (2006) The shape of urban experience: A reevaluation of Lynch's five elements, *Environment and Planning B: Planning and Design*, 33, pp. 803–823.

Stevens, Q. (2007a) 'Betwixt and between: Building thresholds, liminality and public space, in Franck, K. and Stevens, Q. (eds) *Loose Spaces: Possibility and diversity in urban life* (London: Routledge), pp. 73–95.

Stevens, Q. (2007b) *The Ludic City: Exploring the potential of public spaces* (Oxford: Routledge).

Stevens, Q. and Dovey, K. (2004) Appropriating the spectacle: Play and politics in a leisure landscape, *Journal of Urban Design*, 9(3), pp. 351–365.

Stratford, E. (2002) On the edge: A tale of skaters and urban governance, *Social and Cultural Geography*, 3(2), pp. 193–206.

Strathern, M. (1999) 'What is intellectual property after?', in Law, J. and Hassard, J. (eds) *Actor Network Theory and After* (Oxford: Blackwell), pp. 156–180.

Sudjic, D. (2001) Stand off on the South Bank, *The Observer*, 16 September. Available at: http://observer.guardian.co.uk/review/story/0,,552422,00.html [accessed April 2006].

Thrift, N. (2004) Intensities of feeling: Towards a spatial politics of affect, *Geografisker Annaler*, 86b(1), pp. 57–78.

Virginia Museum of Fine Arts (no date) *Rick Mather: A visionary architect inspired by context* (Virginia Musuem of Fine Arts Media Room). Available at: www.vmfa.mediaroom.state.va.us/architectmatherbio.html [accessed July 2020].

Vivoni, F. (2009) Spots of spatial desire: Skateparks, skateplazas and urban politics, *Journal of Sport and Social Issues*, 33(2), pp. 130–149.

Weinzierl, R. and Muggleton, D. (2003) 'What is "post-subcultural studies" anyway?', in Muggleton, D. and Weinzierl, R. (eds) *The Post Subcultures Reader* (Oxford: Berg), pp. 3–23.

Wilcox, G. (2002) What should happen to the South Bank?, *The Observer*, 20 January. Available at: http://observer.guardian.co.uk/review/story/0,,636443,00.html [accessed April 2006].

Wrigley, N. and Lowe, M.S. (2002) *Reading Retail* (London: Arnold).

Zukin, S. (1993) *Landscapes of Power: From Detroit to Disneyworld* (Berkeley: University of California Press).

Zukin, S. (1995) 'Whose culture? Whose city?', in LeGates, R. and Stout, F. (eds) *The City Reader* (London: Routledge), pp. 136–146.

6 Building the Taiwanese mobilityscape[1]

An actor-network account of the journey from Bicycle Kingdom to Cycling Paradise

> If Taiwan was not only exporting bicycles, but also exporting bicycle culture, then there's little doubt that they'd be selling even more bicycles around the world. Time for the Bicycle Kingdom to live up to its name.
>
> (Mayne 2013, quoted in Auchapt 2013)

Introduction

Building the argument from the previous chapters, this chapter shifts across the Taiwan Strait to look at the rise of cycling in Taiwan, where the past 15 years has seen a concerted push to promote cycling at a variety of scales. Somewhat differently from the UK and cities like London, Taiwan's adoption of cycling is complicated by the fact that it remains a global centre for bicycle and component manufacture. Thus its adoption of cycling culture is not a strategy geared solely towards maintaining urban productivity, but potentially one with a much broader impact on national industry. Whilst managerial and technical accounts of the reformation of the Taiwanese bike industry are numerous (Brookfield *et al.* 2008; Chen *et al.* 2009; Chu 2009), cultural-economic accounts are rare (cf. Andreae *et al.* 2013). Accordingly in this chapter we emphasise a cultural mode of economic ordering intended to solve problems of capital accumulation by creating a more culturally embedded domestic industry, and by presenting an image of Taipei as a global 'liveable' city and Taiwan as a Cycling Paradise through educating citizens and visitors about new ways of moving.

In line with previous chapters we see this as a process of mobility fixing, but in this chapter we foreground different empirical and theoretical aspects of fixing. From a theoretical viewpoint, drawing upon actor network theory (ANT), we highlight the importance of processes of ordering, translation and alignment to producing cycling subjects and associations between place and mobility. Here we investigate *how* this transformation is occurring, with a focus on the process as an ongoing and precarious accomplishment. Using an ANT-inspired framework, we describe processes of ordering, translation and alignment through which different actors attempt to create associations between previously disconnected people, places, practices and representations. It is this process of ordering – what Çalışkan and Callon (2010) call 'agencement' – that is central to the creation of cycling culture, and ultimately the valorisation of aspects of practice that enable behavioural surplus to be extracted from employees and citizens.

From an empirical viewpoint we are concerned to emphasise the links between place image and the performance of cycling at different scales from the corporation to the nation. Whilst Taipei can be seen through the promotion of cycling to be trying to achieve very similar goals to cities like London in reducing congestion and improving human capital, we emphasise here the symbolic component of this endeavour. Whilst London is well established as a Western global city, Taipei has only recently begun to compete on this stage, thus the projection of a liveable image is arguably more important for it than for more established cities like London. We present the term 'mobilityscape' to conceptualise the ways in which cycling cultures (sporting, touristic and transport) and their attendant landscapes and infrastructures become a form of 'meta-activity' that (re)produces urban and economic identity (Julier 2005: 869). Paraphrasing Julier (2005) we ask, 'how does [mobility] … contribute to providing official or non-official "stories" and "ways of doing things" that, in turn, fashion specific aesthetic outlooks for a place?' (871). However, we also highlight that symbolic associations between Taipei and cycling are fragile and foreground the importance of performance and material relations in strengthening these symbolic associations.

The structure of the chapter is as follows. First, we discuss some of the theoretical antecedents that inform the chapter, most notably place marketing and ANT approaches to ordering. We then go on to discuss why Taiwan, Taipei and the domestic bike industry need to promote themselves as possessing a cycling culture. In the following section of the chapter we discuss different events, materials and practices through which associations are made between cycling, industry and place to present a cycling culture. We then offer some conclusions as to how this links in to broader debates regarding the political economy of cycling, and its implications for the geographies of cycling.

Cultural economy and place image

Cultural economy has long held the view that what informs notions of value and worth overflows what we consider to be the economic realm (Amin and Thrift 2004: xi). The production and accumulation of resources 'has always been a cultural performance' (Amin and Thrift 2004: xii). As Crewe (2016) has argued, the recognition of cultural economy suggests that value is created in those realms deemed non-economic, and 'problematizes the notion that geographies of production and "Made In" appellations are the key locus of value-creation' (Crewe 2016: 513). As such, culture is seen to play a central role in shaping economic fields in ways that, whilst less measureable and quantifiable, are no less powerful.

Power and Jansson (2008) argue that value (for both firms and places) increasingly resides in the high levels of design, symbolism and aesthetics. However, they also note that these immaterial 'symbolic' aspects of cultural products tend to be difficult to control and stabilise. One way in which such symbolism plays out is in the now familiar process of place branding – a burgeoning activity within advertising and marketing (Olins 1999). According to Julier, the branding of a location can be defined as any attempt 'to create and nurture the narratives that give meaning to a place' (Julier 2005: 872). Whilst our contribution in this chapter is not directly to the place marketing literature, it is a useful conceptual tool as it refers to 'the development of brands for geographical locations such as regions, cities or communities, usually with the aim of triggering positive associations and distinguishing the place from others … by forging particular emotional and psychological associations with a place' (Eshuis *et al.* 2014: 154–155, in Vuignier 2016: 9). Place branding, then, refers to the creation of an identity through activities designed 'to establish a significant and differentiated presence in the market' (Vuignier 2016: 5). Zenker and Braun note that place brand does not refer to the physical, concrete characteristics of a place, but to the perception of these things. It works through 'visual, verbal and behavioural expression of a place', often embodied in the 'general culture' of the citizens to create a network of associations in the consumer's mind (Zenker and Braun 2010, in Vuignier 2016: 35).

Using examples of products like Swiss watches and Belgian chocolates, Kavaratzis and Ashworth (2005) discuss the process of geographic co-branding to describe the relationship between products and the place-bound symbolic qualities of a geographic region where a product is produced (cited in Andersson 2014: 151). Such products are positioned as better than their competitors' because of their geographical origin. Through values such as perceived quality and origin, brands are unavoidably imbued to varying degrees and in differing ways by spatial connections and meanings: 'Commodities are inescapably entangled in geographical knowledges employed in order to re-enchant and differentiate commodities' (Pike 2011: 208).

Whilst there is a significant literature looking at the ways in which landscapes and events are used to re-enchant commodities (Pike 2009, 2011), everyday culture has seen much less attention. The phenomenon of 'event mania', for example, is by now well researched in the literature (Arnaud and Soldo 2015), from festivals and mega-events to medium and small-scale events (Pinson 2016: 835; Vuignier 2016: 33). Cycling is no stranger to the mega-event, with the Tour de France representing the most obvious means through which landscape and event are brought together to showcase the nation. More recently Koch (2018) has demonstrated the ways in which road cycling has been used to enhance place image in other contexts. She illustrates how events such as the 2016 UCI World Race Championships in Doha represent 'an exercise wherein nation-building, place branding, and urban boosterism all intersect' (Koch 2018: 3) as part of a strategy to shape

Doha as modern and developed (Koch 2018: 9). Whilst not always successful, in such instances the image of place is overtly and tightly strategized, as Julier (2005) and others have demonstrated.

Whilst mega-events are of course important, we would argue alongside Pinson (2016) and Julier (2005) that smaller events and everyday culture are also worthy of study in this regard. As this suggests, place image is also formed in less spectacular, less coordinated and less obvious ways. One of the earliest studies in place branding was that of Molotch (1996), exploring 'L.A. as design product'. Whilst recognising the difficulties inherent in defining a singular design aesthetic in a heterogeneous city like Los Angeles, Molotoch was at pains to demonstrate 'how local aesthetics are important to the business climate or "industrial atmosphere"' (Molotoch 1996, in Julier 2005: 873). Molotoch foregrounds the presence of a (often unofficial/unrecognised) relationship between a broader cultural landscape and local industries. As this attests, it is most likely that local aesthetics coalesce from both intentionally and unintentionally networked activities (Julier 2005: 870).

Beyond the heavily contrived and concentrated activities that contribute to place identity, there are evidently more mundane materials and activities that contribute to a more diffuse symbolic landscape.[2] Julier (2000, 2005), for example, coins the term 'designscape' to conceptualise 'the network of activities and artefacts that produce place-identity within cities' (Julier 2005: 869). What Julier foregrounds with this term are the ways in which particular design cultures become a form of 'meta-activity' that frames and reinforces urban and economic identity. As this suggests, this meta-activity may only be a 'loosely coherent' network (Julier 2005: 869).

Julier goes on to critique the focus on iconic buildings, landscapes and landmarks as defining place identity in much place branding literature, arguing instead that a wide range of unofficial activities and everyday artefacts – 'a loosely coherent, hegemonic network of signifiers' – are performed and mobilised in the formation of aesthetic capital (Julier 2005: 869). Importantly, this identity-forming process works not only at the level of broader images to be consumed, but through production and consumption of practices that serve to reconfigure lifestyles and tastes (Julier 2005: 874). Referring to the designscape, Julier notes that it is constituted as much by 'the "things" that make up this fabric as their mediation and symbolic representation' (Julier 2005: 869):

> here, the meaning of nation comes into life through the language and rhetoric that articulate it. This narration is not abstractly independent of objects, however. Instead, what is said is activated by objects, while narration also contextualises and gives meaning to them. Design objects and discourse are mutually dependent in generating belonging.
>
> (Julier 2005: 871)

In such a reading, the practised and mobile can take on the same significance as the material and immobile in the shaping of place image and identity.

Whilst not concerned directly with place branding, Soderstrom (2014) offers an account that foregrounds the mobility of fixed infrastructures (such as shopping malls, road junctions and pedestrianisation schemes) in the process of creating national narratives. For Soderstrom these things act both as 'long-distance government tools to create business-friendly urban milieus' and 'occasions for self-government and learning' (Soderstrom 2014: 145). As Soderstrom argues:

> the power of urban forms resides not only in their capacity to discipline their users, but also in the fact that they enable them to practice the city differently and to position themselves in society by experiencing and expressing in words or deeds new senses of themselves.
>
> (Soderstrom 2014: 146)

Ultimately Soderstrom demonstrates how 'transnational investments, policies, and architectural forms shape new urban practices and imaginaries' that seek to project a particular idea of nation and place (Soderstrom 2014: 145).

Julier asks the question, 'how is design mobilised to affect ... both the individual and collective dispositions and practices that structure cultural capital?' (Julier 2005: 870). Similarly, in this chapter we seek to synthesise and build upon the work of Julier and Soderstrom by expanding our study from the material to encompass the performative within this notion of 'travelling types'. Here we focus on different kinds of orderings and agencements 'to identify the varied impulses and articulations through which value is formed, added and circulated' (Amin and Thrift 2004: xv).

It is a central argument of this chapter that in the Taiwanese context outlined here, the process of mobility fixing works through the construction of a mobilityscape which is itself reliant on the economisation of everyday cycling practice through the interaction of disparate elements. Callon (2007) argues that matters of market concern (such as a downturn in bike sales or place image) produce new entities and identities that are primarily social. As such, the emergence of new, or commodification of existing, communities of practice can be seen as a crucial feature of innovation dynamics. As Callon goes on, 'these communities combine tacit and codified knowledge, group together humans and non-humans (technical devices play a crucial part), and are deployed through a multiplicity of temporal and spatial frames' (Callon 2007: 153). The mobilityscape produced through these practices, whilst at best only loosely strategised, is the result of ordering or 'agencement' (Çalışkan and Callon 2009; Franklin 2004) that seeks to create associations between particular people, objects, symbols and landscapes.

It is with some trepidation that we contribute another 'scape' to an already burgeoning list, which, as Fagence (2014) points out, currently runs to well over 100 (servicescape, designscape, eventscape, festivalscape, sensescape...) and does not always help in conceptualising the processes being explored. This reticence notwithstanding, following Bell (2003) and Van der Duim (2007), in this chapter we present the term 'mobilityscape' to conceptualise the complex processes of ordering through which diverse events, infrastructures, actors and activities come together to negotiate, reconstruct and transmit a particular aesthetic of mobility.

The need for the mobilityscape: Taiwan's industrial, urban and national image

The first question we ask in this chapter is why? Why are different actors in Taiwan concerned to shift mobility towards a cycling culture and in so doing attempt to create associations between Taiwan and cycling? There are a number of reasons why this might be the case, some of which overlap with other cities discussed in the book. Taiwan, however, is also globally significant as a centre of bicycle manufacture, which, as we will elaborate in this section, generates an additional matter of concern.

The Taiwanese bicycle industry grew rapidly from the 1970s, consolidating throughout the 1980s with an increasing global presence. For many years now it has been a hub of global bicycle and component manufacture: two of the world's biggest brands – Giant and Merida – are based there, alongside hundreds of original equipment manufacturer (OEM) suppliers manufacturing for all of the world's largest cycle brands. According to Taitra, there are now approximately 700 companies in the Taiwanese bicycle industry, clustering largely in the Taichung-Changhwa area (Chen *et al.* 2014: 5). However, exports declined in the late 1990s due to global recession, with problems further compounded in the early 2000s as many manufacturers sought to take advantage of lower production costs by moving facilities to China. The net result was a reduction in export units from around 10 million in the early 1990s to under 4 million in 2003 (Chen *et al.* 2014: 5).

Recognising the problem of decreasing sales and margins, many firms in the industry have sought to reposition themselves in the higher-priced end of the market, and to shift from OEM to higher value-added brand manufacturing (Chen *et al.* 2014: 5; Henderson 2003: 37). Two of these companies – Giant and Merida – have successfully transitioned to become two of the world's leading bicycle brands, with, respectively, 10% and 4% of global market share for high-end bicycles and 50% and 25% of the high-end Chinese market (Chen *et al.* 2014: 3). However, a key problem for Taiwan has been the relative lack of a domestic market for its products: Taiwan's bicycle manufacturers generally export 80% of their products, with very little enthusiasm for cycling in Taiwan itself (Chen *et al.* 2009: 209). Chu (2009) confirms this, stating that for Giant, local sales make up less than 10% of total sales (1063).

In an attempt to turn its ailing fortunes around and ensure bicycle production would not leave Taiwan completely, the cross-industry 'A Team', principally comprising Giant, Merida and KMC, was established in 2002[3]

(Brookfield *et al.* 2008: 15). The A Team has sought to reinvigorate the domestic bike industry principally in three main areas: production efficiency, product development, and market intelligence.[4] However, some in the industry recognised that efficiency and innovation alone would not enable a long-term reinvigoration of their fortunes and that something more was required to enhance brand reputation and justify higher prices.

Speaking in 2015, Anthony Lo, Giant CEO, implied that the change in strategy was a recognition of the importance of cycling culture: 'as a bicycle industry we are proud that we are also exporting the cycling experience today [...]. Taiwan converted from cycling island to cycling paradise. Only ten years ago people called Taiwan the cycling desert' (quoted in Van Schaik 2015). Here Lo explicitly links the importance of positioning Taiwan as a place where people enjoy cycling – as opposed to a cycling desert – to the success of its bicycle industry. The danger that Lo highlights is that the 'cultural desert [is] associated with imitation' (Henderson 2003: 41). Hence, to distance oneself from the cultural desert requires not only making bikes, but *doing* bikes. That is to say, industry actors in the A Team realised that to pursue cycling as a purely economic project (e.g. of industrial efficiency) was no longer enough; it must also be pursued as a cultural project.

In particular, as Henderson suggests, good design is linked to cultural proximity, where the 'country of origin' mark on imported goods functions as a kind of 'brand label' in that we have preconceived ideas of product qualities depending on where a product is made (Henderson 2003: 41). However, it is not only where it's made that is important, but, more importantly, who is designing it, and whether they are deemed to 'know' the practice associated with the product. This is exemplified in the image projected by North American and European brand bicycle manufacturers, where firms are represented as being populated by people passionate about riding bikes. Many mid-level mountain bike firms based in California, for example, like Ibis, Ellsworth and Santa Cruz, are testament to this – all (or so the story goes) started by cycle enthusiasts for cyclists, as the founder of Santa Cruz states: 'a bunch of bike geeks making what we believe to be the best products available ... we've got some really fast guys working here' (Davis and Rogers 2012: 109). Even the larger bicycle brands, such as Specialized, SRAM and FOX, whilst much more corporate in their outlook, were started by enthusiasts in a way that (I presume) investment banks never were. Despite their size, for example, Specialized's HQ in the USA still has a lunchtime bike ride for employees and a 'pump' track on site (Privateer 2013), suggesting that its employees perform and embody the values of the company through the practice of cycling: 'you only have to watch the lunch ride roll out or check out the pump track to see how passionate everyone is about bikes' (79).

The result of employees (particularly designers and those in customer-facing roles) not riding bikes is, it is argued, poor design and low-value products. Phil Latz (editor of Australian *Bicycling Trade* magazine), for example, used to write a column in the magazine entitled 'What Were They Thinking?', which outed the worst designs found in any given year: 'I used to see some of the most awful bikes, shockingly bad designs, impossible-to-ride geometries, solutions looking for a problem, and so forth' (Latz, in Brown 2016). However, Latz goes on to say that he ended the column because he no longer had enough material. Asked why the poor designs dried up, Latz suggested it was because 'people in Taiwan's cycling industry decided to actually start riding their own products, which led to a massive improvement in quality' (Latz, in Brown 2016).

Importantly, this distinction between companies that *do* cycling and those that don't is also spatialised and extended to the region and its citizens more broadly. The founder of Santa Cruz bikes, for example, stated that, 'from a credibility standpoint, people take us more seriously because we're in Santa Cruz' (Privateer 2013: 87). In contrast, Merida's marketing manager felt that Taiwan as a place contributes less symbolic capital to their brand: 'in Taiwan, traditionally our position in the marketing image is lower than say California for example' (Merida Head of Marketing, Interview, May 2017). Certainly the Eastern (non-cycling) and Western (cycling) images of the employee/citizen as possessing different characteristics constitutes a dissociation that many in the Taiwanese industry believe needs to be overcome. Hence the divisions here are not only between employees who do or do not ride, but are extended to whether the region, in general, has a cycling culture. As this suggests, to build a cycling culture, and to enrol one's employees within it, is seen as a way of enhancing brand value.

Fixing national and city image through cycling culture

Taiwan is not known for its cycle culture: numbers of cyclists, whether recreational, tourists or commuters, have been low to non-existent. In 2011 less than 5% of the population owned a bicycle, with only 4.6% riding with any regularity, a figure virtually unchanged by 2016. In the capital Taipei, the mode share of cycling is little better, with 5.1% using the bike in 2015 compared with 26% using motorcycles. This absence of cycling culture has increasingly been seen as a problem for Taipei, in particular when seeking to compete on the global stage. Seeking to position itself as a global, liveable and progressive city, it has looked to learn from the example of select Western cities where cycling is more prevalent.

In Chapters 2 and 4 I emphasised the bio-political and transport-economic facets of the mobility fix, where cycling serves to improve the functioning of the city through improved circulation times and human capital (see also Spinney 2016). Here we emphasise the symbolic place image facet of fixing which serves to solve problems of accumulation through image-building. As discussed in Chapter 2, according to Harvey (1989), the role of particular fixes in urban governance has 'become much more oriented to the provision of a "good business climate" and to the construction of all sorts of lures to bring capital into town' (Harvey 1989: 11). Thus Harvey has posited a shift from managerial forms of urban government to entrepreneurial forms of governance once distinctive to private commerce (Gulsrud *et al.* 2013: 330), as cities must increasingly prove their competitive worth on a global stage.

Cities such as Vancouver, Melbourne, Copenhagen and Vienna consistently vie for the accolade of 'most liveable city', a title granted by various business, human resource and lifestyle publications (see, for example, Monocle, Economist Intelligence Unit). In such liveability rankings, a high position indicates an 'internationally business-friendly city and thus attracts new global commerce and can boost local pride in a city's reputation' (Gulsrud *et al.* 2013: 331). Liveability in particular emphasises 'green city brands [which] have become a global tool for municipal leaders to promise a better quality of life, promote sustainable development, and increase their competitive advantage' (Gulsrud *et al.* 2013: 330).

Within such urban entrepreneurialism, mobility and cycling specifically are playing an increasingly important role. Jennings (2015) points to the link between cycling and city liveability in the context of public bike-sharing schemes. She cites the example of Cape Town, where the city strategy explicitly points to the link between inter-city competition and green transport (12–13). Indeed, she goes on to note that the ITDP's Bike Share Planning Guide explicity suggests 'that a public bicycle system could "improve a city's image and branding"' (Jennings 2015: 13). Similarly she notes that when establishing a case for a PBSS in New York, the ITDP foreground the 'city image' and media benefits of public bicycles (Jennings 2015: 13). We argue here that Taiwan is attempting to create a similar set of associations. Numerous interviewees in government, NGOs and business noted the potential impact of cycling culture for enhancing the global image of the city: 'Through promoting green transport, we would step by step make Taipei City one of the world-renowned Green capitals' (Secretariat of Taipei City Government (TCG) 2014: 10). Indeed, according to one official, it was the wish to project the image of a green city to a global audience that led to the You Bike pilot (PBSS) scheme launching in Taipei's most globally projected and visited district 'Xinyi' (Taipei City): 'many foreigners travel frequently to Xinyi District', which is 'the embodiment of a global city' (Dr Li, Former Commissioner of TCG's Department of Transportation, Interview, 19 October 2015).

Direct comparisons were drawn by key officials between the perceived change of image from implementing PBSS in other cities:

> We observed from the bike program in Paris that they promoted public bikes and bike lanes, and then the authority saw the change in the numbers of bike usage and in the city image.... So if you examine the YouBike project in Xinyi District, you will see that the ways they

> set up bike stations and the charge rate are exactly the same with the bike program in Paris.
>
> (Mr Shi, YouBike Manager in TCG's Department of Transportation, Interview, 23 October 2015)

Indeed, the same interviewee goes on to discuss the limited impact of YouBike on Taipei's traffic: 'YouBikes are replacing pedestrians … YouBikes never ever replace motor vehicles,' adding that 'we can feel that YouBike is a showcase, just for show; it doesn't have any real effect on transportation.' Hence what emerges here is the idea of PBSS – and cycling more broadly – as a strategy of urban entrepreneurialism – of 'display' – as much as, if not more than, a transport solution. However, it is also important to recognise the post-colonial nature of such imaginaries where 'not-so-global cities' are forced to play the 'bigger game of winning some kind of world recognition', and pursue an 'art of being global that is invariably caught up in a political game that is allusive, contrastive, comparative, and contested, with cities in the region, but also beyond' (Ong 2011: 13).

Clearly it is not only Taiwanese industry that is seeking to reinvigorate its image through adopting cycling: with city and national image in mind as well as industrial productivity, disparate stakeholders, including Merida, Giant, the Taipei city government, Ministry of Tourism and Communication, Ministry of Transport, and Taiwan Cycling Federation, amongst others, have all contributed to a strategy attempting to inculcate cycling culture in the population more broadly in order to enhance the image of place and product as progressive. In the following sections we discuss the associations and translations that have been made in an attempt to assemble this mobilityscape. These include a temporally and spatially diverse range of performances and materials that constitute the mobilityscape, including events such as the Tour de Taiwan cycle race, King of the Mountains Challenge, Chunghua Classic 100, Taiwan Bike Festival, Taipei International Bike Show, Velo-City 2016 Cycle Conference, Bike Taiwan Cycle Route, and Taipei's 'YouBike' PBSS. Some of these, such as the Velo-City Conference, are one-off events (2016), whilst others, such as YouBike, are 'permanent'. Equally, some events, such as the International Bike Show, take place only in Taipei city, whilst others, such as the Bike Taiwan Cycle Route, are evident in all areas of the island. Importantly, we demonstrate that in order to realise the symbolic value of cycling in the cause of image-building, cycling subjects are brought into being in the firm and the nation through these events in order to perform cycling.

International study visits

The association of Taiwan with cycling as a social practice does not just magically happen: work must be done to align Taiwan/Taipei/the cycle industry with the doing of cycling. The former mayor of Taipei, for example, referred to the importance of associations with other (cycling) cities to initiating the process of association:

> In 2011, the mayor of Bern (Switzerland), accompanied by our officials of the Ministry of Foreign Affairs, came to visit me. He said that Taiwan was world-renowned for creating a cycling paradise, yet he wondered why he couldn't see the cycling trend in Taipei city.
>
> (King Liu, Giant Bikes Founder, cited in *Business Weekly* 2015)

Through conversations with government officials, the international study visit emerges as a key translation through which spatial alignment is accomplished. According to Vicky Liu (Giant Bike Foundation), many Taiwanese central and local government officials have been invited on different visits in order to understand what cycling in Taiwan could look like: 'I always say to others that you must visit the Netherlands for inspection … you may not be able to totally adopt their approaches to promoting cycling, yet you will learn the logic and principles behind their approaches.' In this instance, the international study visit functions to create alignment between Taiwan and a well-known cycling nation, the Netherlands. As Liu points out, this is not an exact translation, as it is acknowledged that Taiwan 'cannot totally adopt their approaches'. However, it still seeks to import the 'logic and principles' of a European urbanism by facilitating the bicycle with a corresponding reduction in car use (Tønnesen 2015) and therefore make the idea of Taiwan as a cycling nation comprehensible.

This idea was reinforced by other stakeholders who used numerical projections derived from study visits to position Taiwan on the same trajectories as other pro-cycling cities:

> The cycling cultures in those 'developed countries and cities' mentioned in the inspection report such as Netherlands and Frankfurt were far better than that in Taiwan. The number of bike trips occupies 25% out of all traffic trips in Netherlands and 10% in Frankfurt (Wei-yu Chu *et al.* 2012), while only 3.9 to 5.5% in Taipei.
>
> (Zhu *et al.* 2012)

A similar dissonance was echoed by another interviewee: 'Mode share for bikes in Taipei has grown from 1.3% in 2003 to 5% this year – about 350,000 trips per day. We expect to reach 12% by 2020. Total YouBike trips has already exceeded 26 million' (TCG DoT 2015). Whilst the figures themselves suggest a distance between the cycling cultures of these places, through these projections Taiwan attempts to create alignment with cycling cities and nations, not by equalling their cycling levels, but by narrating that it is on the same trajectory and therefore comparable.

Conferences

Having competed against numerous other cities, Taipei won the hosting rights to the 'Velo-City Global' conference held between 27 February and 1 March 2016. The conference is the largest cycle-related conference in the world, regularly attracting around 1,500 academics and industry professionals from 65 different countries (ECF 2017). The hosting of the conference is important primarily because of its role in associating Taiwan, and Taipei specifically, with a broader cycling culture.

The first way it does this is spatial and symbolic, by positioning Taipei alongside other cycling cities that have hosted the conference, including Barcelona, Vienna and Copenhagen. The very fact that Taipei 'won' the right to host the conference legitimises it as a cycling city and seeks to create equivalence between Taiwan and cycling. Indeed, the conference internet homepage broadcasts this legitimacy with recourse to its winning the bid to host the conference: 'Taipei was selected over Changwon, Korea, and cities in China and India to host the 2016 edition, clear affirmation of the city's energetic efforts in promoting bicycling' (Taipei City Government 2016). We emphasise two processes of translation and ordering here: the first is one that seeks to create equivalence between Taipei and other cycling cities through the common denominator of having hosted Velo-City. The second is one that seeks to disassociate Taiwan from 'non-cycling' nations – Korea, India, China – nations that have not hosted Velo-City. In this instance, whether a city or country can be associated with cycling or not is linked to whether it has hosted the conference or not.

The second way in which the conference serves to associate Taiwan and cycling is by acting as a medium to perform the association between the nation, its cities, industry and cycling culture. The main mode through which this was achieved was discursively through different presentations and conference themes throughout the four days of the conference. In all of the conference themes, Taiwan conveyed its involvement in different aspects of cycling (manufacture, design, bike-share, tourism, cycling infrastructure, sport, commuting…), which, when taken together, might constitute a cycling culture. In order to convey this holism, the conference organisers were at pains to ensure that all the key plenary sessions included presenters from international cities known to be cycle-friendly alongside government officials from Taiwan and Taipei, and representatives from its cycle industry. This is exemplified by the plenary on 'Sustainable Urban Transport'. This session featured officials from cities all highly positioned in liveability rankings: Stephen Yarwood (former mayor of Adelaide); Maria Vassilakou (deputy mayor of Vienna); Morten Kabell (Copenhagen Kommune); alongside Charles Lin (deputy mayor of Taipei) (Velo-City field notes, 27 February 2016). This 'billing' represents a process of alignment in that it attempts to present Taipei as being 'on the same page' as these other established cycling cities by being on the same stage. Again, the translation functions by attempting to create equivalence through proximity and association between geographically and culturally distant places.

Cinematic events

Other more diffuse events are also crucial in establishing temporal and spatial associations. Directed by Chen Huai-En, *Island Etude* was Taiwan's entry to the 80th Academy Awards in the category of Foreign Language Film. *Island Etude* is effectively a road movie where the main teenage character goes on a journey of (cultural, geographical, social, personal) discovery and enlightenment around Taiwan by bike.

Island Etude was cited by numerous interviewees as a key actant in the promotion of cycling culture in Taiwan. Lin, from the Taiwan Cycling Federation (TCF), stated, for example, that the film was responsible for an increased interest in recreational cycling from 2007–2010 because it positioned cycling as more than poor person's transport: it suggested that successful people could ride too (Mrs Lin, TCF, Interview, January 2015). Perhaps more importantly, King Liu (founder of Giant bikes) has also gone on record as stating that the film catalysed his inspirational first navigation of the island by bike:

> 'With some things, if you don't do them now, you might never do them in your life,' was its most often repeated line. 'I saw the film and thought, that line, it's criticizing me, isn't it?' Mr. Liu said. 'I was 73, and I thought if I don't ride now, I'll never be able to do it.' So he took up the film's challenge, planning a circumnavigation of Taiwan.
>
> (Ramzy 2013)

Island Etude is important in the process of agencement because it attempts to align Taiwan with cycling by presenting its landscapes and culture in direct relation to cycling. It is also no coincidence that the week-long round-the-island bike tour presented in the film is very similar to that performed by King Liu, and promoted in key tourist documents since, enabling its performance by both citizens and tourists alike (Stratton 2018).

Alongside *Island Etude*, other broadcast representations have also emerged. The documentary series *Discover Taiwan*, for example, aired on the UK Bike Channel throughout 2017. This multi-part documentary follows the journeys of British TV presenter Anna Johnson as she makes her way around the island by bike. The film makes associations between cycling and Taiwan's heritage through constantly collaging images of temples, traditional rituals and cycling alongside 'traditional' music. Similarly, each episode of the documentary travels (along the official round-the-island bicycle network) through a different county on the island, taking in nationally significant landscapes and structures. The resulting image is one that again, much like *Island Etude*, attempts to make the idea of Taiwan as possessing a cycling culture comprehensible by creating associations between culture, place and cycling.

Mass participation events

Taiwan hosts a growing series of cycle-related events happening throughout the year, aimed at both professional and recreational cyclists. Promotion of these began in 1999, when the new sport-focused Taiwan Cycling Federation (TCF) began to cooperate with government on the promotion of a series of events: family rides; cycle round island; university cycling event (Mrs Lin, TCF, Interview, January 2015). Whilst these started off as stand-alone events (for example the King of the Mountain (KOM) events sponsored by Merida, the professional road race Tour de Taiwan staged in March, Formosa 900 sponsored by Giant), since around 2010 the TCF has cooperated with the Department for Tourism and Communication to bring many of them together under the badge of the 'Taiwan Cycling Festival' to give greater coherence (Mrs Lin, TCF, Interview, January 2015).

In 2015, for example, the Festival ran from 30 October to 22 November (double the length of time from the previous years) centred around four main events: 'Taiwan KOM Challenge'; 'Formosa 900'; 'Sun Moon Lake Come Bikeday'; and 'OK Taiwan Bike and Horse Riding Tour'. This programme was complemented with cycling activities in ten different counties. According to the Ministry of Tourism and Communication (MoTC), the main purposes of the festival are:

> to make international tourists 'think of Taiwan in seeing bikes' as well as 'thinking of a trip to Taiwan in seeing bikes', and furthermore to reach the three major goals of Taiwan Cycling Festival: 'being internationally renowned', 'promoting the cycling tourism industry' and 'LOHAS of the whole people'.
>
> (MoTC 2015)

These events establish equivalence and align disparate entities in a number of ways. First, they are used to weave a temporal narrative of cycling history. The internet homepage for Velo-City 2016 on the Taipei City Government website explicitly establishes a cycling past for Taiwan through these events:

> large-scale activities promoting cycling for sport and exercise have been repeatedly staged. Among them is Taipei Cycle (Taipei International Cycle Show), first held in 1988, and into the 26th edition this year. Another major event is the Tour de Taiwan, established in 1978 and run 22 times to date, which in 2006 was integrated with Taipei Cycle.
>
> (Taipei City Government 2016)

By giving the date established and number of times the events have run, such statements establish a cycling past for Taiwan. This acts to create equivalence with other cycling nations where the presence of a cycling culture is in part announced through a calendar of well-established cycling events (for example European nations such as France, Belgium, Spain and Italy that host a full season of pro and amateur cycling events).

The popularity of these cycling events in Taiwan has increased greatly in the last ten years, as the marketing manager of Merida noted in relation to the Chunghua Classic 100 km event. Merida has been the main sponsor for the event since 2010, with the number of riders rising from 300 in the first year to 8,000 in 2017. Entry is open to all abilities but gives amateurs the chance to ride alongside professionals:[5]

> Now when we announce the date, normally within one day all the spaces have gone. Also, many of them are foreigners so they see it as an important event with amateur teams organising together and coming to Taiwan to enjoy that race.
>
> (Merida Head of Marketing, Interview, May 2017)

Events such as the Tour de Taiwan, Chunghua Classic and KOM establish equivalence through associating Taiwan with broader signifiers of cycling culture. This is accomplished discursively and performatively. Discursively, events like the 'King of the Mountains' borrow from the history of established cycling events as one of the jerseys that can be won in all the major cycling Grand Tours (Tour de France, Giro D'Italia, Vuelta a España). In using this name, Taiwan is positioning its event alongside these internationally reknowned cycling events. Beyond this, the landscapes through which the courses are routed also enable participants to perform classic cycling landscapes. The KOM event climbs up to 3,275 m above sea level (the highest point on the island accessible by road); the Chunghua Classic takes in 670 m of climbing. Both events aim to position Taiwan's mountain landscapes alongside European Alpine climbs made famous in cycling circles through their inclusion on the Grand Tours and other smaller 'classic' races. These events serve to align Taipei's mountains with a broader road racing cycling culture of Alpine ascent (Spinney 2006). As such, they represent a 'dramatic ritualization of symbols and values' deploying 'verbal and behavioural rhetoric to assert power in uncertain systems of meaning and representation' (Arnould and Price 2000: 148).

The trade show

A further set of translations and alignments we discuss are those made between Taiwan's bike industry and its emerging cycling culture. The key point we emphasise here is the redefinition of the bike industry as something more than just an economic entity: it becomes entangled in culture through specific events and relations, as elaborated upon by Merida's head of marketing:

> Lots of companies, not just Merida, we try to tell to the outside world that we are not just a manufacturing island for the bicycle, it's also a very good island for cycling. The idea probably started from ten years ago because we know that to create a culture, we will also eventually benefit our local sales in Taiwan.
>
> (Merida Head of Marketing, Interview, May 2017)

Taiwan hosts an annual international cycle show (TICS) in the Xinyi district of Taipei. It is the third largest in the world behind Interbike (Las Vegas) and Eurobike (Friedrichshafen), although Shanghai's cycle show is emerging as a competitor to Taipei (Andreae *et al.* 2013: 197). The show is primarily staged to highlight technology, innovation, and facilitate relationship-building rather than direct sales (ibid.). Given that the 'trade show is a societal space used by many exhibitors to establish trust in their trading relationships' (Skov 2006, in Andreae *et al.* 2013: 194), we would expect the product – cycling – to be embodied and performed by employees at these shows. Yet Phil Latz, editor of Australian *Bicycling Trade* magazine, paints a different picture of employees at past TICS: 'The show was dominated by suit-and-tie wearing chain smokers … the hall was filled with smoke, and the people in Taiwan's cycling industry … well, they didn't seem like the sort of people who cared much about cyclists' (Latz, in Brown 2016). Benedict confirms this view of the non-cycling employee:

> it seems obvious that the top brass in any cycling company should be an avid cyclist … [yet] many of the largest [cycle] companies are simply run as businesses, where the widgets could be interchangeable with any other widget as far as many executives are concerned.
>
> (Benedict 2012)

The dissociative image of cycle industry employees in this account is that of the suit-wearing, golfing, smoking executive who has no real interest in the product they are selling (beyond the profit that can be made), and whose bodily comportment sits at odds with a normative Western European and American image of the cycle employee as active and healthy and passionate about the product being made. This reading was confirmed by a member of the leading Taiwanese cycling advocacy organisation, the Taiwan Cyclist Federation. She noted that when she worked for Giant in the 1980s, 'all the executives played golf and were not interested in cycling', partly because of Giant's predominantly inward OEM focus (Mrs Lin, TCF, Interview, January 2015). Similarly, a senior cycling advocate and former Giant employee noted that 'the time when we were at Giant, we always suggested our bosses to ride our own bicycles to promote, yet they were just playing golf ... they considered that cycling is the activity of the lower class at that time' (Mr Su, TCF, Interview, January 2015).

However, there was evidence of a substantial shift away from golfing, smoking, suited cycling executives toward employees embodying cycling in our observations at the 2016 Taipei International Cycle Show:

> The main floor of the cycle show is laid out with the big international brands dominating the central area using imposing and glossy stands. These are surrounded with somewhat smaller and less opulent stands from smaller brand manufacturers and some aspiring OEMs. The perimeter of the main area is made up of small uniform booths occupied almost exclusively by small Taiwanese and Chinese OEM and original design manufacturer (ODM) suppliers. The staff on the brand manufacturer stands 'look the part', many of them dressed in cycling clothing but all appear sporty, relatively young and constantly moving round the stands. The staff in the OEM booths are much more mixed in their appearance: many of them are sitting down, often looking bored (perhaps due to their peripheral position) and generally not wearing cycle clothing. However, when I strike up conversations, the majority seem to be very knowledgeable about the product, with many professing their love of (most commonly road) cycling.
>
> (field diary, March 2016)

According to Brunner and Dever (2014), the new labour processes represented here are bound up in broader trends which blur work and personal identities in service economies (in Good and Cooper 2016: 451). Certainly, Nickson *et al.* (2001) have argued that 'maintaining an appropriate look and disposition is also a key part of the labour performed, designed to fit the organisation's image, improve the customer experience and increase sales' (Nickson *et al.* 2001, in Good and Cooper 2016: 451). Evidently the idea that employees perform an appropriate image of cycling at the trade show is central to creating associations between making and doing that seek to enhance the value of the products on display. Here the appearance and knowledge of exhibitors reflects on the quality of products because it suggests embeddedness in the social practice of cycling, rather than just the process of manufacture. Thus it is through the alignment and ordering of bodies, spaces (such as the trade show) and narratives that symbolic value is created.

Cycling employees

A number of interviewees discussed a shift towards cycling amongst employees in the bike industry: 'The shift of Giant Manufacturing's attitude toward bicycles started from the European branch's suggestion' (Mrs Hu, Chair of Taiwan Cyclist Federation, Interview, 10 December 2015).

This story of a shift in participation was corroborated by Lin at the Taiwan Cyclist Federation. When asked why she thought cycling culture had become important for Giant, she suggested that it was 'because Giant now sees the bike with a global view' (Mrs Lin, TCF, Interview, January 2015). In particular, Lin cites the importance of King Liu (founder and president of Giant) starting to cycle, as it signalled to suppliers how important it was to cycle. The head of marketing at Merida corroborated this, stating that, because Liu was and is an 'opinion leader' in the industry, it meant that his story became central to selling cycling within the industry (Merida Head of Marketing, Interview, May 2017).

Such was the change that from 2007 King Liu required 'all employees in Giant's department of domestic sales, divided into several groups to travel around the island by bike by the end of the year' (Bike Taiwan 2007). This requirement of employees to ride was also echoed by Chung Moon-Cheung, vice president of Merida:

> We encourage employees to ride around the island to punch in because the mileage is also regarded as one of the regulations to employees ... If our employees have a healthy body, it is a great help for our company to promote new projects ... have a common topic can also improve relationships between colleagues.
>
> (Chung Moon-Cheung, quoted in *Manager Today* 2012)

One established strand of research exploring the economic importance of cultural performance is that exploring the ways in which the performances of employees are managed and shaped to contribute to the success of the firm. Writing in 1983, Hothschild demonstrated that the development of the service sector had led to the emotions and feelings of employees becoming organisational commodities. Hothschild demonstrated that the act of presenting and regulating a particular emotional register became caught up in 'making a sale' and argued accordingly that emotions must be seen to have exchange value (Hothschild 1983, in Watson and Ward 2013: 2905). Similarly, Crang (1994) has studied the ways in which labour performances in a restaurant are carefully dictated, surveilled and managed by employers. More recently, Good and Cooper (2016: 451) have examined sexuality as a strategy in the workplace, 'where the organisation mandates a particular employee look for commercial benefit, often drawing on the sexuality of employees' (Warhurst and Nickson 2009). In the accounts we present here, it is the mobility of employees that is mandated for commercial benefit. Much like the associations witnessed at the trade show, the mass cycle events already discussed (such as the island circumnavigation) act as a vehicle through which employees can learn what it means to cycle, and in doing so perform the brand in a way that (it is hoped) adds value. Thus these events provide a forum to simultaneously entrain a cycling sensibility and present it to suppliers.

Merida's head of marketing, for example, discussed at length the importance of such events in presenting an image of cycling culture to foreign buyers:

> these events are very big and normally we invite our foreign customer. [...] We invite these people to come to Taiwan to try the event. In the meantime they can see our factories, they can see our concept store and they can see the event we organise. Sometimes they also invite their dealers in Taiwan. Oh, such good factories! They go to the shop in Taipei: oh such a big shop! They come to Taiwan, we keep telling them how good Merida is and they can see it. They can see the real shop, see the people, see the nice mountain bike or road event, and they will try it and believe it, and they will tell their friend and customer: Merida is good, Taiwan is good: buy from Taiwan no mistake.
>
> (Merida Head of Marketing, Interview, May 2017)

What the executive emphasises here is the importance of participation to creating successful association. He goes on to say, 'it's important that they can come, they can see it by themselves, and they can touch, they can feel the temperature of the company. If they can really trust then they believe' (ibid.). The idea surfaces again that creating successful association is dependent upon a greater immersion than simply reading or seeing – convincing associations are premised upon touch, feel and participation, as this executive went on:

> its important that they come back to the home market and get a feeling that it's a right decision for them to work with Merida: Taiwan is right as a home because they can also ride. Then they see our factory and our shops and people. It is so important to show them not just the manufacturing side, but some other soft side about cycling culture … they know cycling in Taiwan is important, not just for production.
>
> (Merida Head of Marketing, Interview, May 2017)

Hence mass cycling events represent a process of agencement through which different actors (citizens, employees, sales executives and foreign buyers) are aligned in order to perform a cycling culture that it is hoped will enhance trust in the firm.

National cycle network and tourism

Taiwan is engaged in ongoing attempts to position the island as a destination for bicycle tourists, both domestic and foreign. Speaking at the Velo-City conference in 2016, Monica Kuo, speaking on behalf of the Ministry of Tourism, gave a talk promoting the creation of a green bike network across the island to encourage tourism and benefit industry (field diary,

1 March 2016). The rationale for providing this cycle-friendly landscape was reinforced by the Central News Agency (2015), which stated that the promotion of cycling was intended 'to create ... the better value of tourism ... cycling has gradually become the mainstream of the domestic tourism and recreation. It may connect the local hot springs and gourmet food and bring huge benefits to tourism.' This position is echoed in reports by the Department of NGO International Affairs (DNGOIA):

> It is expected that the around the island bicycle trails will become better and better and Taiwan will be turned into a 'cyclists' paradise' so that more and more foreign cyclists will be able to ride bicycles and explore the beauty of Formosa.
>
> (Dai-Ru 2017)

In order to enable both citizens and foreigners to participate in cycling, it is recognised that Taiwan cannot rely on isolated events and symbolism alone: as Van der Duim (2007) has stated, associations that don't hold together on their own must be performed. As Law puts it, there is 'no such thing as a purely social relation' because materials are constantly delegated to enact relations (2001: 4). What becomes evident in the Taiwanese context are the ways in which materials and spaces are increasingly utilised to translate cycling from something that could be done to something that is done. These are most notably the production of a bicycle network across the island and within Taipei city, and the development of the 'YouBike' PBSS in Taipei city (amongst others).

Cycling is first mentioned as a strategic tourism priority in 2009 but became much more pronounced from 2011, with cycling positioned as part of a strategy for: 'Promotion of healthy travel, development of Green Island and Little Liuqiu as ecotourism demonstration islands, and creation of new spotlights in tourism development' (Tourism Bureau, MoTC 2017). Development of Taiwan as a cycle-friendly island was evidently on the agenda prior to this, however, even if it was not presented as a strategic priority. The most notable manifestation of this is found in plans to develop a 940 km network of regional bikeways between 2003 and 2007.[6] This was to be overseen by the National Council for Physical Fitness and Sports at an estimated cost of NT$1.517 billion. The plan maps out nine regional networks of bike paths 'to form "green corridors" linking different areas together, with the gradual building up of local networks and the interconnection of around-the-island and regional networks, to provide a complete path system for recreational biking' (MoTC 2008: 34–35). Moreover, the route would specifically incorporate elements of 'high architecture' (such as bridges and ancient buildings) and natural landscapes in both coastal and foothill settings[7] (Monica Kuo, Velo-City field diary, 1 March 2016). After completion, work would be carried out to join these up to form a national network, and also to integrate them with rail travel (MoTC 2008: 34–35).

The main benefits of this were stated as the formation of 'a complete network for green recreation and tourism, integrating recreation and sports facilities as well as main tourist spots around the country, spurring regional and local development, and delivering an economic boost at the grassroots level' (MoTC 2008: 34–35). Much of the inspiration for this network has been from the successful Shimanami Kaido cycle route in Japan. As Mr Hsu from the Ministry of Transport confirmed, numerous government departments (including education, tourism and transport) alongside businesses like Giant have been involved in defining the route around the island (Mr Hsu, Ministry of Transport, Interview, May 2017).

By 2008, progress on the network had been significant, with detailed design guidance and plans drafted for much of the network; construction of 600 km of bike paths in Taipei, Kaoshung, Taichung and Hsinchu cities and seven counties through a system of grants; and provision of vital inter-regional and topographical links through the building of a series of bridges.[8] Development has continued on the network: according to the Ministry of Tourism and Communication, the final 'mainline' links of the network would be in place by November 2015 and all branch lines connected by the end of 2018 (Shan 2015: 5).

The creation of the network also enrolled a number of other actants in performing cycle tourism. As the DNGOIA stated, alongside the route itself they also set up:

> better infrastructure elements such as road signs, delimiting lines, marking of mileage, supply stations, and tourist guiding information boards. Currently the police stations, schools, and retail shops along the bike trails form the 'supply chains' which provides water and tire-pumping for cyclists.
>
> (Dai-Ru 2017)

The associative significance of this cycle network for the practice of cycling is both symbolic and practical. At the symbolic level it aligns Taiwan alongside other nations where cycling enjoys special status and is supported by significant investment, such as the Netherlands and Denmark. Similarly to other associations, the development of the national and regional cycle networks works through equalisation – the idea that Taiwan has a bikeway network that sets it on a par with nations that have established cycling cultures. The network also functions at an experiential level; if cycle tourism (and in particular the island circumnavigation) were promoted without the network, it would arguably be unsuccessful as cyclists would find themselves on dangerous roads or unable to travel between specific places. As a result, the proposed symbolic association of Taiwan with cycling would fall apart. The materiality of the cycle network, from the tarmac to the signage, increases the likelihood of the successful performance of the promoted association between Taiwan and cycling. Again, this represents a process of translation because the network shrinks the gap between representation and practice: it makes the idea of cycling and the practice of cycling more equivalent (though not completely equal) and in so doing makes the idea of Taiwanese cycling culture a comprehensible idea.

Taipei city – PBSS and bike lanes

King Liu, founder of Giant Bikes, publicly stated that 'Taipei, with its car and motorbike-clogged streets, is the biggest obstacle to making Taiwan a "cycling island"' (Ramzy 2013). This was also recognised by city officials, and in 2007 the then mayor of Taipei city confirmed the policy of 'cycling in daily life' and asked the Department of Transportation (DoT) to create a relaxed and safe cycling environment:

> In 2008, the DoT under the Taipei city government delegated Giant manufacturing to work on the pilot project for establishing public shuttling bike system and operations management. Giant bid NT$268 million for the procurement on a 7-year contract. They chose Xinyi planning district to be the pilot zone and built public bike stations. On 11 March 2009, a pilot project was launched with 11 docking stations and 500 public bikes.
>
> (Department of Transportation 2009)

The role of YouBike in positioning Taipei as a modern and global 'liveable' city cannot be underestimated, with numerous interviewees articulating the idea that in conveying a cycling culture, PBSS associates Taiwan with other modern (Western) cities. Speaking at the Velo-City (2016) conference, Vicky Yang from YouBike stated that 'we always check if they have a PBSS when we go to cities – we are proud to be on the list of cities that has a PBSS.' To reinforce this, she shows a slide from Boris Johnson (former mayor of London) stating that there are 'two kinds of cities: one with PBSS and the other planning PBSS' (field diary, Velo-City, February 2016).

PBSS in Taipei in particular is positioned as a way of creating an urban landscape that mirrors the European city by clearing out undesirable scooters and replacing them with YouBikes: 'what will it be like if there are no scooters parking in the pavement in Taipei? I believe it will be like a European city' (Mr Po, Former Commissioner TCG Department of Transportation, Interview, 25 November 2015). This narrative attempts to associate the performance of commuting with bikes and disassociate it from scooter riding. Indeed, one of the main government reports argued that bikes and scooters were equivalent: 'The solution offered by this report is to introduce bicycles to replace scooters instead of cars … because of the similarities between cyclists and scooter riders in the aspects of purpose of travel, distance, age groups and so on' (Department of Transportation Management 2008: 45). Consequently, the city government began to reallocate space and facilities from scooter riders to bike riders. According to one official, scooter riders 'suffered unspeakably from the project' with '5,326 scooter parking spaces cancelled and lane widths reduced due to the establishment of the grid cycling network' (Huang *et al.* 2016). Similarly, pedestrians were asked to share space with cyclists: 'as public bikes became quite popular, we … allowed bikes to ride on all the pavements with pavement width above two meters … Soon it provoked controversy … about affecting the current walking space for pedestrians' (Mr Lo, Former PBSS Manager in Taipei Department of Transportation, Interview, 23 October 2015).

Further clues regarding the importance of PBSS in Taipei to the emergent mobilityscape are given by interviewees discussing the genesis of the system. This was most evident with interviewees discussing PBSS:

> If you want to make Taiwan a cycling paradise … you have to jump out of the frame of current transportation modes and reconsider how to have everyone become cyclists and how to encourage people to choose bikes as their major means of transportation.
>
> (Mrs Lo, CEO of YouBike, Interview, 14 November 2015)

> YouBike learnt from international experiences: looked to Paris Velib to build PBSS in itself, and to Netherlands to understand integration of bike with other modes … YouBike in Xinyi District set up bike stations and charge rate exactly the same as the PBSS in Paris.
>
> (Mr Lo, Former PBSS Manager in Taipei Department of Transportation, Interview, 23 October 2015)

Mr Lo went on to explicitly state that 'PBSS in Taipei City is modelled on European cities: Amsterdam, Copenhagen, Lyon, Paris' (Mr Lo, Former PBSS Manager in Taipei Department of Transportation, Interview, 23 October 2015).

These statements represent a process of alignment and ordering where Taipei is positioned alongside other cycling cities – Amsterdam, Copenhagen, Paris. As the marketing manager of Merida summarised:

> I believe that PBSS contributes to the image of Cycling Paradise. We have similar YouBike system in 5 or 6 counties so I think that does help: consumer or foreigner come to Taiwan and they can see that [PBSS] and that can help to turn us into cycling island or even better cycling paradise so I think PBSS can contribute a lot.
>
> (Merida Head of Marketing, Interview, May 2017)

As this suggests, PBSS serves to project the image of a cycling city to a global and domestic audience. Similarly to the National Bike Network, the materiality of the system also increases the chances of the association of

Taipei as a cycling city being successful by facilitating its performance. Indeed, as Vuignier states:

> it is no longer enough to simply 'sell' a place; instead, one must alter and improve it in accordance with a long-term vision and expectations, using a marketing logic involving interaction between current and potential target groups and the facilities, residents and organisations already on site.
>
> (Vuignier 2016: 13)

Trouble in Cycling Paradise – dissociation

As Julier has observed, 'while clear, easy-to-understand narratives of a place may be easy to develop and communicate in themselves, they also require confirmation in the material circumstances to which they relate' (Julier 2005: 872). A successful narrative – in this case the everyday practice of cycling – requires citizen participation. Performance of the narrative is a process of association: if the narrative is not performed, or a counter-narrative is performed, the desired associations are unlikely to take place and the ultimate goal will not be achieved. This was recognised by interviewees: 'YouBike ... wants to change scooter riders and drivers, making them give up riding their own scooters and cars and take public transportation' (Ms Fon, YouBike Manager, Interview, 10 January 2016). Our interviews with those involved in cycle promotion in Taipei suggest that for a variety of reasons the associations and dissociations required for Cycling Paradise to be a success were partial at best because of the difficulty in translating the narrative into spaces and practices.

In previous sections we have discussed how authorities attempted to spatialise cycling culture by implementing cycle lanes and PBSS, primarily in a central grid system: 'The "three latitudinal and three longitudinal" bicycle network was the significant project of the Department of Transportation of Taipei City for hosting the Velo-City Global 2016' (Huang *et al.* 2016). However, as many interviewees suggested, the attempt to shift people from scooter culture to cycling culture was unsuccessful in the city context: 'scooter riders and drivers are still riding scooters and driving cars' (Mr Lo, Former PBSS Manager in Taipei Department of Transportation, Interview, 23 October 2015). Indeed, despite attempts to encourage people to drop their scooter and ride to work or school, the reality at least domestically is that the scooter is still the preferred mode of travel in Taipei.

In particular scooter riders complained that space had been taken from them to build lanes for YouBikers but 'no one rode on the bike lanes, so why it occupied one traffic lane? That was the point that drivers and scooter riders often attacked' (Mr Lo, Former PBSS Manager in Taipei Department of Transportation, Interview, 23 October 2015).

> Scooter riders' protest against the Dunhua bike lanes was the most crucial (factor) to the cancellation of the bike lanes … They felt a sense of deprivation; that is, there were not so many cyclists, yet why were there exclusive bike lanes yet no exclusive scooter lanes?
>
> (Prof. Lam, Policy Consultant for TCG Department of Transportation, Interview, 2 November 2016)

As a result of this backlash, plans to build dedicated bike lanes were rolled back and instead, YouBikers were enabled to share space with pedestrians: 'Cyclists will be allowed to ride on the sidewalks with cycling road signs. "Pedestrians First" will be added on the bottom of the cycling road sign' (Taipei City Police Department 2017). The result of this was twofold: first, from a practical point of view, cyclists were less likely to perform the cycling culture because it was less safe and convenient than had a fully separated set of lanes been implemented: 'The biggest problem of riding YouBikes is "no bike lanes for YouBikes"; cyclists are "either hitting pedestrians on sidewalks or getting hit by cars on roads"' (Mayor Ko, cited in *Liberty Times* 2014). Second, from the perspective of projecting an image, the decision to make cyclists share space with pedestrians sends a signal that cyclists are still not important. If a key element of cycling culture is support from society more generally, this suggests that Taipei has some way to go.

Conclusions

There are a number of conclusions we draw from this chapter, contributing both empirically and theoretically. First, this chapter represents a further empirically grounded case study demonstrating how and why cycling is manifesting in particular locations in specific ways. As we have shown, the ways in which cycling touches down in the context of Taiwan and Taipei are in many ways similar to other nations and cities attempting to bring cycling back from the margins. Through a series of events, narratives, archetypes and materials similar to those found in many other countries, the version of cycling manifested in Taiwan draws upon a standardised repertoire of global imaginaries and policy mobilities (Soderstrom 2014). However, driven by a somewhat different agenda – Taiwan, for example, is in part seeking to boost the industrial image of its domestic cycle industry as embedded in cycling – Taiwan brings a different range of actors to bear in both assembling and performing the image of cycling.

Second, we have presented this process of image-building in Taiwan as a process of ordering and aligning that attempts to create cycling subjects both within the cycling industry and the population at large. This endeavour is vital to the success of the resulting mobilityscape because the associations that constitute it are prone to fail if they are not performed.

Thus any successful image requires actors, props and sets that can facilitate the desired performances. As we have shown in this chapter, the process of ordering seeks to enrol study visits, conferences, films, mass participation events, trade shows, bike-sharing schemes, and tourism infrastructures, amongst others, to align Taiwan with cycling as a practice. It is through these orderings that the social practice of cycling is both brought into being and economised. This, we argue, represents a shift in the mode of ordering from economic to cultural economic. Whereas initial attempts to reinvigorate industry and tourism in Taiwan have been focused on efficiency and productivity (Brookfield *et al.* 2008; Chen *et al.* 2009; Chu 2009; Lee and Huang 2014), we demonstrate here a shift towards embedding economy within culture in order to create value. Evidently both industrial and urban entrepreneurialism increasingly work through culture. As Law (2007) has argued, the shift in mode of ordering changes the possibilities for connection and association, in this instance enabling value to be created by connecting place, industry, mobility and culture.

Finally, this chapter has argued that this process of image-building exemplifies another aspect of mobility fixing. Whilst Chapter 4 emphasised the contribution of mobility to fixing problems of congestion and sedentary human capital, in addition this chapter has foregrounded the work of fixing in a symbolic register. Indeed, in the case of Taipei and Taiwan, we argue that the creation of cycling subjects is as much an attempt to create and project an image of place as liveable and Western, and industry as embedded in practice, as it is an attempt to change mobility patterns within the city and nation. To precis the words of one city official in relation to Taipei's YouBike scheme: PBSS has changed mobility very little but provides an excellent showcase for Taipei and Taiwan's bicycle industry on the global stage. As such, image-building is seen to rely on the production and extraction of behavioural surplus at the population level. Whilst employees and citizens are encouraged to cycle to work, the symbolism of their aggregate performance is harnessed in service to other urban and national economic goals.

Ultimately this chapter highlights the importance of cultures of mobility to reinvigorating capital accumulation. We have demonstrated how temporary events and activities and permanent infrastructures and places are mobilised to produce a diffuse mobilityscape. By attempting to associate and align what were previously detached, we demonstrate the importance of mobility to understanding contemporary processes of capital accumulation. In particular we emphasise that it is through projecting a new image of movement rather than (for example) an iconic skyline or mega-events that Taiwan seeks to transform its place identity and industrial image. We have also highlighted the precarious nature of this undertaking, demonstrating its emergence as an ongoing accomplishment prone to failure and always threatening to collapse.

Notes

1 This chapter is based on a previously published paper co-authored with Dr Wen-I Lin at National Taipei University: https://journals.sagepub.com/doi/abs/10.1068/a43149
2 Whilst the etymology of landscape is complex, the English use of the word has its roots in sixteenth-century painting and refers to the representation of natural scenery. In human geography more specifically, landscape tends to be viewed as a way of seeing – the suggestion in both is that landscape refers to the development of a set of unifying features that create distinction within a given territory or region.
3 Early statistics from 2005 suggested some success, with A Team members seeing growth of 14% in exports; average unit prices increasing to $346; and sales volumes up by an average of 36% (Brookfield *et al.* 2008: 16).
4 Of most relevance here is the A Team's marketing operations embodied in the 'Taiwan Industry Image Enhancement Project' (IEP). This aimed 'to enhance the innovative image of Taiwanese industry, creating the country of origin effect and assisting Taiwanese branded companies in raising brand awareness and consumer preference, as well as increasing sales revenue and expanding the sales base in the targeted markets, including in Taiwan, the EU, USA, Japan, Mainland China, India, Indonesia, Vietnam, Philippines, Russia, Brazil, UAE, Egypt, Myanmar, Turkey, and Mexico' (www.taiwanexcellence.com.tw/ind/about-us.aspx, accessed 25 April 2017). Supported by the Ministry of Economic Affairs and implemented by the Taiwan External Trade Development Council (TAITRA), the IEP provides value-added promotion through a variety of integrated marketing and communication activities (conferences, exhibitions, digital marketing, etc.) (ibid.).
5 The 2017 edition, for example, gave entrants the chance to ride alongside Pro-Team professionals Valerio Agnoli and C.K. Feng.
6 Much of the inspiration for this network has been from the successful Shimanami Kaido cycle route in Japan. As an official from the Ministry of Transport confirmed, numerous government departments (including education, tourism and transport) alongside businesses like Giant have been involved in defining the route around the island (Mr Hsu, Ministry of Transport, Interview, May 2017).
7 This route was inspired by the Japanese Shimanami Kaido route (part administered by Giant, who provide bike rentals), which runs 60 km across a series of islands in the Seto Sea connecting the main island of Honshu with Shikoku.
8 Grants were provided for bike paths across Huajhong Bridge, Fuhe Bridge, Yongfu Bridge, Taipei Bridge, Guangfu Bridge and Huajiang Bridge (MoTC 2008: 30).

References

Amin, A. and Thrift, N. (eds) (2004) *The Cultural Economy Reader* (Oxford: Blackwell).

Andersson, I. (2014) Placing place branding: An analysis of an emerging research field in human geography, *Geografisk Tidsskrift-Danish Journal of Geography*, 114(2), pp. 143–155.

Andreae, M., Hsu, J-Y. and Norcliffe, G. (2013) Performing the trade show: The case of the Taipei International Cycle Show, *Geoforum*, 49, pp. 193–201.

Arnaud, C. and Soldo, E. (2015) Le portefeuille territorial d'évènements culturels (PTEC): Nouvelle modalité de gestion de l'offre d'évènementiel culturel pour un management stratégique des territoires, *Management International / International Management / Gestión Internacional*, 19(2), pp. 115–127. doi: 10.7202/1030390ar.

Arnould, E. and Price, L. (2000) 'Authenticating acts and authoritative performances: Questing for self and community', in Ratneshwar, S., Mick, D.G. and Huffman, C. (eds) *The Why of Consumption: Contemporary perspectives on consumer motives, goals and desires* (London: Routledge), pp. 140–162.

Auchapt, D. (2013) *Taiwan Needs to Become a Cycling Island: ECF at the Taipei Cycle Show* (European Cyclists Federation). Available at: https://ecf.com/news-and-events/news/taiwan-needs-become-cycling-island-ecf-tapei-cycle-show [accessed 9 May 2017].

Bell, D. (2003) Mythscapes: Memory, mythology and national identity, *British Journal of Sociology*, 54(1), pp. 63–81.

Benedict, T. (2012) Taiwan bike tour: Giant Bicycles Taichung factory visit, *BikeRumour*. Available at: www.bikerumor.com/2012/10/18/taiwan-bike-tour-giant-bicycles-taichung-factory-visit/ [accessed 10 November 2017].

Bike Taiwan (2007) *Don't Waste a Moment of Life: King Liu accepts the round Taiwan bike challenge*. Available at: www.wheelgiant.com.tw/publication/bmuen/html/bmuen200703/BMUEN200701_P068.pdf [accessed August 2018].

Brookfield, J., Liu, R-J. and MacDuffie, J.P. (2008) Taiwan's bicycle industry A-Team battles Chinese competition with innovation and cooperation, *Strategy and Leadership*, 36(1), pp. 14–19.

Brown, J.S. (2016) Taiwan: from Bike Kingdom to Cyclist's Paradise, *Taiwan Business*. Available at: https://topics.amcham.com.tw/2016/04/taiwan-bike-kingdom-cyclists-paradise/ [accessed April 2016].

Brunner, L. and Dever, M. (2014) Work, bodies and boundaries: Talking sexual harassment in the new economy, *Gender, Work and Organization*, 21(5), pp. 459–471.

Business Weekly (2015) The first comes for the unique: Life thinking questions shared by Giant's King Liu, *Business Weekly* (Taipei: Cite Publishing), pp. 157–158.

Çalışkan, K. and Callon, M. (2010) Economization, part 2: A research programme for the study of markets, *Economy and Society*, 39(1), pp. 1–32.

Callon, M. (2007) An essay on the growing contribution of economic markets to the proliferation of the social, *Theory, Culture and Society*, 24(7–8), pp. 139–163.

Central News Agency (CNA) (2015) *Welcome to 2016 Bicycle City Conference by over 500 Friendly Shops* (Taipei: CNA). Available at: http://goo.gl/jOj86L [accessed October 2016].

Chen, J., Shieh, C. and Yamada, S. (2014) Taiwan bicycle sector, *Credit Suisse Securities Research and Analytics*.

Chen, Y-S., Lin, M-J., Chang, C-H. and Liu, F-M. (2009) Technological innovations and industry clustering in the bicycle industry in Taiwan, *Technology in Society*, 31, pp. 207–217.

Chu, W. (2009) Can Taiwan's second movers upgrade via branding? *Research Policy*, 38, pp. 1054–1065.

Crang, P. (1994) It's showtime: On the workplace geographies of display in a restaurant in southeast England, *Environment and Planning D: Society and Space*, 12, pp. 675–704.

Crewe, L. (2016) Placing fashion: Art, space, display and the building of luxury fashion markets through retail design, *Progress in Human Geography*, 40(4), pp. 511–529.

Dai-Ru, L. (2017) *Around-the-Island Bicycle Trails Will Cause Another Cycling Boom in Taiwan* (Department of NGO International Affairs). Available at: www.taiwanngo.tw/p/16-1000-26991.php?Lang=en [accessed November 2017].

Davis, M. and Rogers, S. (2012) Striking a balance, *Privateer*, 10(July), pp. 95–110.

Department of Transportation (2009) *Annual Report of 2008* (Taipei: Department of Transportation).

Department of Transportation Management, Tamkang University (2008) *Taipei City Bicycle Policy Research* (Taipei: Research and Development Evaluation Committee of TCG).

Eshuis, J., Klijn, E-H. and Braun, E. (2014) Place marketing and citizen participation: Branding as strategy to address the emotional dimension of policy making?, *International Review of Administrative Sciences*, 80(1), pp. 151–171.

European Cyclists Federation (2017) *The Global Cycling Summit*. Available at: https://ecf.com/projects/velo-city [accessed August 2018].

Fagence, M. (2014) 'Scape'-based forms: A preliminary review of their use in the study of tourism-related activities, *Tourism Recreation Research*, 39(2), pp. 275–283.

Franklin, A. (2004) Tourism as an ordering, *Tourist Studies*, 4(3), pp. 277–301.

Good, L. and Cooper, R. (2016) 'But it's your job to be friendly': Employees coping with and contesting sexual harassment from customers in the service sector, *Gender, Work and Organization*, 23(5), pp. 447–469.

Gulsrud, N., Gooding, S. and Konijnendijk van den Bosch, C. (2013) Green space branding in Denmark in an era of neoliberal governance, *Urban Forestry and Urban Greening*, 12, pp. 330–337.

Harvey, D. (1989) From managerialism to entrepreneurialism: The transformation in urban governance in late capitalism, *Geografiska Annaler*, 71B(1), pp. 3–17.

Henderson, S. (2003) Designed in Taiwan, *Design Management Journal*, 14(2), pp. 36–41.

Huang *et al*. (2016) The councillor was angry, citizens need to wait for 90-second red traffic lights seven times: three horizontal and three vertical cycling lane policy makes Taipei fall into Traffic darkness, *China Times Newsletter*. Available at: http://goo.gl/gyqny8 [accessed October 2016].

Jennings, G. (2015) Finding our balance: Considering the opportunities for public bicycle systems in Cape Town, South Africa, *Research in Transportation Business and Management*, 15, pp. 6–14.

Julier, G. (2000) *The Culture of Design* (London: Sage).

Julier, G. (2005) Urban designscapes and the production of aesthetic consent, *Urban Studies*, 42(5/6), pp. 869–887.

Kavaratzis, M. and Ashworth, G.J. (2005) City branding: An effective assertion of identity or a transitory marketing trick?, *Tijdschrift voor economische en sociale geografie*, 96(5), pp. 506–514.

Koch, N. (2018) The geopolitics of sport beyond soft power: Event ethnography and the 2016 Cycling World Championships in Qatar, *Sport in Society*, 21, pp. 2010–2031. doi:10.1080/17430437.2018.1487403.

Law, J. (2001) *Ordering and Obduracy* (Lancaster: Centre for Science Studies, Lancaster University). Available at: www.comp.lancs.ac.uk/sociology/papers/Law-Ordering-and-Obduracy.pdf.

Law, J. (2007) *Actor Network Theory and Material Semiotics*. Available at: www.heterogeneities.net/publications/Law2007ANTandMaterialSemiotics.pdf [accessed 30 June 2017].

Lee, C-F. and Huang, H-I (2014) The attractiveness of Taiwan as a bicycle tourism destination: A supply-side approach, *Asia Pacific Journal of Tourism Research*, 19(3), pp. 273–299.

Liberty Times (2014) Mayor Ko: The biggest problem of YouBike is running public bikes without bike lanes, *Liberty Times*. Available at: http://goo.gl/Cxd3xs [accessed October 2016].

Manager Today (2012) *Merida: Chung Moon-Cheung: Not afraid of difficulties, drive the organization to challenge high goals*, trans. Wen-I Lin. Available at: www.managertoday.com.tw/articles/view/15148 [accessed November 2017].

Ministry of Transportation and Communication (2008) *Challenge 2008: National Development Plan (2002–2007)*. Available at: www.cepd.gov.tw/2008/challenge2008.pdf [accessed November 2017].

Ministry of Tourism and Communication (Taiwan) (2015) *Annual Report on Tourism 2015*. Available at: http://admin.taiwan.net.tw/upload/contentFile/auser/b/annual_2015_htm/English/ [accessed August 2018].

Molotch, H. (1996) 'L.A. as design product: How art works in a regional economy', in Scott, A.J. and Soja, E.W. (eds) *The City: Los Angeles and urban theory at the end of the twentieth century* (Berkeley: University of California Press), pp. 225–275.

Nickson, D., Warhurst, C., Witz, A. and Cullen, A.M. (2001) 'The importance of being aesthetic: Work, employment and service organisation', in Sturdy, A., Grugulis, I. and Willmott, H. (eds) *Customer Service: Empowerment and entrapment* (Basingstoke: Palgrave).

Olins, W. (1999) *Trading Identities: Why countries and companies are taking on each others' roles* (London: The Foreign Policy Centre).

Ong, A. (2011) 'Introduction: Worlding cities, or the art of being global', in Roy, A. and Ong, A. (eds) *Worlding Cities: Asian experiments and the art of being global* (Chichester: Blackwell Publishing), pp. 1–26.

Pike, A. (2009) Geographies of brands and branding, *Progress in Human Geography*, 33(5), pp. 619–645.

Pike, A. (2011) Placing brands and branding: A socio-spatial biography of Newcastle Brown Ale, *Transactions of the Institute of British Geographers*, 36, pp. 206–222.

Pinson, J. (2016) From the Olympic dream to a down to earth approach: Lausanne's sports events hosting strategy, *Sport in Society*, 19(6), pp. 828–839.

Power, D. and Jannson, J. (2008) 'Outside in: Peripheral cultural industries and global markets', in Baerenholdt, J.O. and Granas, B. (eds) *Mobility and Place* (Farnham: Ashgate), pp. 167–178.

Privateer (2013) Santa Cruz part one, *Privateer*, 16, pp. 81–89.

Ramzy, A. (2013) A maker of bikes now makes a point of riding them, *New York Times*, 30 August. Available at: https://nyti.ms/12TQwZA [accessed 9 July 2020].

Secretariat of Taipei City Government (TCG) (2014) *A Report of Mayor-led Delegation to Singapore to Participate in the World Cities Summit Mayors Forum* (Taipei: TCG).

Shan, S. (2015) Cycle route around entire nation set to open in November, *Taipei Times*. Available at: www.taipeitimes.com/News/taiwan/print/2015/06/06/2003620070 [accessed August 2017].

Skov, L. (2006) The role of trade fairs in the global fashion industry, *Current Sociology*, 54, pp. 764–783. doi:10.1177/0011392106066815.

Soderstrom, O. (2014) *Cities in Relation: Trajectories of urban development in Hanoi and Ouagadougou* (Oxford: Wiley-Blackwell).

Spinney, J. (2006) A place of sense: A kinaesthetic ethnography of cyclists on Mt Ventoux, *Environment and Planning D: Society and Space*, 24(5), pp. 709–732.

Spinney, J. (2016) Fixing mobility in the neoliberal city: Cycling policy and practice in London as a mode of political–economic and biopolitical governance, *Annals of the American Association of Geographers*, 106(2), pp. 450–458.

Stratton, M. (2018) *Destination Taiwan: Taiwan by bike, around the island in 12 days*. Available at: https://edition.cnn.com/travel/article/taiwan-cycle-tour/index.html [accessed August 2019].

Taipei City Government (2016) *VeloCity Global*. Available at: http://english.gov.taipei/ct.asp?xItem=77765409&ctNode=8557&mp=100002 [accessed May 2017].

Taipei City Government Department of Transportation (2015) *Road Network of Taipei City* (Taipei: Taipei City Government). Available at: http://goo.gl/As7l3p [accessed October 2016].

Taipei City Police Department (2017) *Traffic Police Brigade*. Available at: https://td.police.gov.taipei/News_Content.aspx?n=9EABCAAA75983B75&s=CD8B89FD9691CCE5 [accessed September 2017].

Tonnesen, A. (2015) Policy packages and state engagement: Comparing car-use reduction policy in two Norwegian cities, *Journal of Transport Geography*, 46, pp. 89–98.

Tourism Bureau, MoTC (2017) *Tourism Policy* (Ministry of Tourism and Communication). Available at: http://admin.taiwan.net.tw/public/public_en.aspx?no=6 [accessed 25 June 2017].

Van Der Duim, R. (2007) Tourismscapes: An actor-network perspective, *Annals of Tourism Research*, 34(4), pp. 961–976. doi:10.1016/j.annals.2007.05.008.

Van Schaik, W. (2015) Taiwan bicycle export expands in value and volume, *Bike Europe*, 3 November 2015. Available at: www.bike-eu.com/sales-trends/nieuws/2015/11/taiwan-bicycle-export-expands-in-value-and-volume-10125044 [accessed May 2017].

Vuignier, R. (2016) *Place Marketing and Place Branding: A systematic (and tentatively exhaustive) literature review*, University of Lausanne Working Paper de l'IDHEAP 5/2016 (Unite Management public et marketing, May).

Warhurst, C. and Nickson, D. (2009) 'Who's got the look?' Emotional, aesthetic and sexualized labour in interactive services, *Gender, Work and Organization*, 16(3), pp. 385–404.

Watson, A. and Ward, J. (2013) Creating the right 'vibe': Emotional labour and musical performance in the recording studio, *Environment and Planning A*, 45, pp. 2904–2918.

Zenker, S. and Braun, E. (2010) *Branding a City: A conceptual approach for place branding and place brand management*, Paper presented at the 39th European Marketing Academy Conference, 1–4 June.

Zhu, W.Y., *et al.* (2012) Study on green vehicles around MRT station: Taking the Netherlands, Paris and Frankfurt as examples, *Rapid Transit Systems and Technology*, 46, pp. 87–104.

7 Transport solution or vehicle for surveillance capitalism?[1]

A case study of dockless public bike-sharing (PBSS2.0) in Shanghai

Introduction

The intersections between physical and virtual mobility are providing fertile ground for new ways of accessing mobility to emerge (Bardhi and Eckhart 2012; Glotz-Richter 2016; ITF 2019). In the case of forms of access such as public bike-sharing systems (PBSSs), these new opportunities are also grounded and made possible by the existing associations between cycling and sharing. Cycling has for some time been lauded for being environmentally sustainable, space-efficient and healthy; digitally enabled sharing economies have been promoted as more pro-social and pro-environmental ways to organise consumption (Hall and Ince 2017; Richardson 2015). Put the two together in the form of public bike-sharing systems (Stromberg 2017; DeMaio 2009; Spinney and Lin 2018) and surely you have a system beyond reproach and critique; a system whose planet and citizen-saving credentials are beyond question?

Whilst PBSSs are rapidly evolving and evidently can have a role to play in urban mobility systems, this chapter seeks to bring a much needed critical perspective to bear. As a result, in some ways this chapter is not about cycling at all, but about the ways in which the bringing together of two forms of mobility – virtual and physical – create the grounds for new avenues of accumulation and surveillance to emerge. Indeed, as Zuboff (2019) argues, applying a broader critical lens to such systems strips away 'the illusion that the networked form has some kind of indigenous moral content, that being "connected" is somehow intrinsically pro-social, innately inclusive, or naturally tending toward the democratisation of knowledge' (Zuboff 2019: 9).

The emergence of public bikes is generally traced back to the 'White Bike' system implemented in Amsterdam in 1965 (DeMaio 2004). However, the 2007 launch of the Velib system in Paris is widely regarded as creating the template for contemporary PBSSs and their rapid global growth (Shaheen *et al.* 2010). In the past ten years, PBSS has exploded in popularity: according to the 'Bike-sharing World Map', there were 1,328 schemes worldwide in operation as of July 2017, with a further 405 planned or under construction (DeMaio and Meddin 2017).

Things took a new turn in early 2015 when a new wave of PBSS2.0[2] that utilises smartphones to lock and unlock bikes – freeing them from the need to have physical docks – emerged in Chinese cities (see, for example, Spinney and Lin 2018). Whilst these systems have had their fair share of bad press, they have also been lauded for their contribution to mitigating urban environmental problems. By way of example, the UN recently awarded Mobike (one of the foremost operators of the new dockless/flexible wave of bike-sharing) with their 'Champions of the Earth' Award (*Bikebiz* 2017b). As Hall and Ince have commented, particularly for those in the world of business, 'these new sharing economies have been perceived … as exciting new innovations promising to transform our economic and social lives' (Hall and Ince 2017: 1). In this chapter we seek a critical understanding of what is driving the emergence of these systems and how they are reshaping social relations. Indeed, following Hall and Ince (2017), we argue that there is a need to consider the ways in which post-2008 capitalism is reinventing itself, and how these formulations intersect and overlap with 'other forms of valuing, exchanging, producing and consuming' (2).

Drawing upon literatures on surveillance/data capitalism and mobilities, we theorise PBSS2.0 as a hybrid apparatus of smartphone and bike deployed to generate behaviour in order to extract surplus value from it. Following Zuboff (2019), we argue that PBSS2.0 forms part of an emerging ecosystem of objects that represents an unprecedented penetration of data capitalism. The evidence we present for this argument is largely circumstantial for two reasons: first, PBSS2.0 companies, like many other tech companies, are very secretive about their operations and hence evidence largely has to be gleaned and deduced regarding their operational black box; second, companies like Mobike and Ofo are in their infancy, and whilst, like Google and Facebook, they may well have been started with pro-social goals in mind, it is still unclear what they will become. Accordingly our argument is that they are on the same trajectory as established surveillance capital firms regarding how they operate and can become profitable, but there is no absolute certainty: at the time of writing all the signs point toward similar outcomes or outright failure.

In line with this, this chapter takes the idea of mobility fixing in a slightly different direction, arguing that PBSS2.0 represents a form of mobility fixing because, in bringing together physical and virtual mobility, new opportunities for accumulation emerge with the potential for already existing or induced behaviours to be rendered productive. In such a reading, the surfeit of mobility produced by unsustainable land-use patterns in the modern city represents a surplus of labour that previously could not be put to productive use. As Jessop (2006) has stated, such surpluses can be absorbed through temporal displacement via investment in long-term (speculative) capital projects; and spatial displacements through opening up of new markets and new production capacities (152). PBSS2.0 is suggestive of both these scenarios because the investment is in an unproven (speculative) business model in that no one knows if the data can be monetised; and it is also evidently geared toward opening up new markets through the creation of new products (i.e. new sets of data that can be sold). As we argue here, this process of valorisation is effectively one of economisation as commodity relations are extended 'into spheres not currently subject to the logic of accumulation' (Jessop 2002: 29).

The structure of the chapter is as follows: we first give an overview of surveillance capitalism, situating the emergence of PBSS2.0 as an articulation of this within the smart city. In the second section we examine the origins and expansion of PBSS2.0 in Shanghai, arguing that its origins in tech companies, funding by both venture capital and internet giants, and rapid expansion all mirror the development of other surveillance capital firms and position it as such. We then go on to discuss the ways in which PBSS2.0 has largely eluded critical examination precisely because it has been linked to ideas of cycling and sharing that share positive associations and obfuscate its unprecedented status. We then move on to look at the apparatus of PBSS2.0, with a particular focus on the ways in which hybridity and movement enable the generation and capture of behavioural surplus. We highlight the ways in which the app interface and points system as well as regulating and encouraging the user also gather valuable data that can be used to create predictive products to be sold on behavioural futures markets. We also highlight the ways in which route data can be used to create advertising opportunities to retailers. In the final section we then go on to discuss the rollback of PBSS2.0, which has withdrawn in most cities as rapidly as it deployed. We then conclude with some observations regarding the ways in which PBSS2.0 changes social, economic and political relations. Most importantly, we emphasise that PBSS2.0 represents another facet of mobility fixing because of the way in which it enables the economisation of social practices such as cycling.

Surveillance capitalism

Behrendt (2016) has argued that 'there is a close relationship between Smart Cities, the Internet of Things and Intelligent Transport as all three concepts rely on the increasing use and experience of networked and sensor technologies' (158). Within this nexus, however, cycling has received very little attention (ibid.) despite – as Schwanen (2015) has noted – the relatively early symbiosis of cycling and apps with regard to self-monitoring (Piwek *et al.* 2015); health tracking (Griffin and Jiao 2015; Mone 2015); and sports (Romanillos *et al.* 2015). As Behrendt comments, these are all examples of the datafication of cycling, where information on use and experience is collected and analysed at scale (Behrendt 2016: 158). In the ensuing section we focus on the work of Zuboff (2019) to provide a framework through which to understand the process and implications of PBSS2.0 as an example of datafication.

Zuboff (2019) defines surveillance capitalism as 'a new economic order that claims human experience as free raw material for hidden commercial practices of extraction, prediction and sales' (8). She persuasively argues that we are entering a new era where surveillance – embedded in our everyday lives – becomes the foundational mechanism in transforming investment into profit: 'under this new regime, the precise moment at which our needs are met is also the precise moment at which our lives are plundered for behavioural data. The result is a perverse amalgam of empowerment inextricably layered with diminishment' (Zuboff 2019: 53).

As French and Smith (2016) highlight, the dawn of surveillance capitalism comes at a time of rapid growth in the personal information economy when a multitude of devices can track and record the minutiae of everyday life conjoined with innovations in the ability to digitally extract, synthesise and visualise bodies and populations (5). Zuboff goes on to break down the logic and operation of surveillance capitalism, stating that data is the raw material necessary for the manufacturing process; extraction describes the social relations and material infrastructure through which the firm asserts authority over those raw materials; and analysis refers to the complex of highly specialised computational 'machine intelligences' that are required to process the raw materials into products with value (Zuboff 2019: 65).

The key data source to be monetised is our behaviour, whether that be web browsing, use of social media, listening practices, or our daily commute. Some aspects of what companies learn from our behaviour are fed into product optimisation or service improvement; the rest is what Zuboff calls 'behavioural surplus' (2019: 8). Behavioural surplus is fed into machinic manufacturing processes and fabricated into 'prediction products' that anticipate 'what you will do now, soon or in the future' (ibid.). These prediction products are traded and exchanged in what Zuboff calls 'behavioural futures markets' (ibid.) to customers, foremost amongst whom are advertisers and financial services (cf. Deville and Van der Velden 2016). Central to increasing the value of these products is the ability to combine data: a prediction product is worth more if it can predict behaviour, wants and needs better, and it is better at doing this if it understands more aspects of our daily lives: what we do, where we go, how we feel, who we relate to (Zuboff 2019: 8). Accordingly, Zuboff highlights the importance of economies of scale (extent of operations), scope (breadth of behaviours observed/knowledge gathered) and action (the active shaping of behaviour through architectures of extraction). As Zuboff is at pains to point out: these processes don't just gather data on behaviour; they actively shape it, representing a shift from knowledge to power (ibid.).

Using Harvey (2005), Zuboff argues that surveillance capitalism represents a form of accumulation by dispossession:

> what accumulation by dispossession does is to release a set of assets … at very low (and in some instances zero) cost. […] entrepreneurs who

> are determined to 'join the system' and enjoy 'the benefits of capital accumulation' are often the ones who drive this process of dispossession into new, undefended territories.
>
> (Harvey 2005: 153, in Zuboff 2019: 99)

Yet evidently there is a cost for the capitalist in that users tend to want some kind of improvement in their lives for handing over their data – often in the form of assistance in achieving particular goals, from exercising to more convenient family logistics – even if they have little idea of what is being handed over or how it will be used.

As this suggests, the success of smart forms of governmentailty relies on the subjectification of citizens as *active* and goal-oriented in order for these technologies to be useful and used (Vanolo 2014: 887, in Krivy 2018: 15–16). As a result, new forms of citizenship are enacted environmentally and emerge gradually through interaction with computational and sensing technologies (Krivy 2018: 15–16). Indeed, as Zuboff has argued, 'dispossession is not a single action but rather an intricate convergence of political, social, administrative, and technical operations … under-pinned by *a predictable sequence of stages … incursion, habituation, adaptation,* and *redirection*' (Zuboff 2019: 136–137). Accordingly, central to the extraction of behavioural surplus is embodied practice in the realm of social reproduction. As demonstrated in Chapter 4, the labour that surveillance capitalism relies upon is the unwaged labour of our everyday lives, from communicating with friends to cycling to work (French and Smith 2016: 7). The ability to produce a behavioural surplus is dependent on 'inducing action rather than restricting it, or, more precisely, by "curating" a networked terrain within which action is nurtured' (Krivy 2018: 19). Hence dispossession is made possible by the provision of particular architectures that direct action in particular ways. For Zuboff, this constitutes a theory of change where goal-oriented subjects are configured through alignment with 'an elaborate range of administrative, technical, and material capabilities' (2019: 136–137).

Into this field we want to introduce dockless public bike-sharing (PBSS2.0). In contrast to its predecessor, where bikes had to be picked up and dropped off at geographically fixed docking stations, PBSS2.0 bikes are located, unlocked and returned using a mobile phone app. This allows much more flexibility for the user with regard to where they start and finish their journey. The key innovation here is that the virtual mobility afforded by the smartphone is brought together with that of the physical mobility afforded by the bike. The result, as Kitchin and Dodge (2011) highlight, is that in imbuing everyday objects 'such as domestic appliances, handheld tools, sporting equipment, medical devices, recreational gadgets, and children's toys, with capacities that allow them to do additional and new types of work', software allows everyday objects to be subject to 'external processes of identification and linkage that embeds them in the emerging "Internet of things"' (Kitchin and Dodge 2011: 47).

Origins and expansion: PBSS2.0 emerges in Shanghai 2015

In this section we first examine the origins and expansion of PBSS2.0, arguing that its origins in tech companies, funding by both venture capital and internet giants, and rapid expansion all mirror the development of other surveillance capital firms and position it as such. The focus of the chapter is on the Chinese mega-city Shanghai. However, other cities such as Guangzhou or Beijing could just as easily have been the focus: the 'choice' as such was more serendipitous than strategic due to a pre-existing research focus on PBSS1.0 in the city at the time.

Shanghai is a city of approximately 25 million residents covering an area of 6,340 km^2. Density of the city varies greatly, with some central districts close to 1,000 p/ha (people per hectare) down to 20 p/ha in some outer areas: average city density is around 200 p/ha (Haixiao 2017: 2). Ownership of motorised vehicles is around 0.50 per resident, with around 20% of journeys in central Shanghai made by car (Haixiao 2017: 30). Modal share for two-wheelers (powered and non-powered) has declined gradually from 17.5% in 2011 to 15.7% in 2015 (ibid.). However, after many years of prioritising cars, bicycles are once again promoted in more recent Shanghai policy documents. A key pillar of this promotion is that from 2009 onwards, PBSS1.0 was instituted in five districts, and by 2017 Shanghai boasted 80,000 government-initiated public bikes, putting it amongst the top five in the world in terms of bike numbers if not usage (see Figure 7.2). Indeed, as reported in other papers (Lin and Spinney 2020 forthcoming), PBSS1.0 in Shanghai has suffered from locational and operational issues, meaning ridership has been relatively low. However, in early 2015 the new wave of PBSS2.0 that utilises smartphones to lock and unlock bikes – freeing them from the need to have physical docks – emerged in Chinese cities, including Shanghai.

The State response to PBSS2.0 in Shanghai

Since 2008 many municipal governments have been suffering budget cuts but are still under pressure to improve their transport systems and encourage green transport. Existing PBSS1.0 systems have required government subsidies, such as London's Santander bikes (with a subsidy of more than £10 million per year) or a private investor (for example an advertising firm) such as New York's PBSS1.0 Phase 2 (Interview with PBSS Manager, TfL, 24 July 2015) in order to provide their services. Whilst Chinese cities have been more insulated from post-2008 cutbacks, it is no surprise that dockless bike-share companies have been welcomed with open arms by many cities (particularly in China and the UK) as a way of providing sustainable transport at (seemingly) zero cost to their taxpayers. As Mr Han Zheng, Government Party Secretary of Shanghai, stated, 'let me make myself clear first: we are for bike-share. We highly approve of such an innovative act'

Figure 7.1 Docked public bikes in Shanghai, 2018.
Source: Author.

Figure 7.2 PBSS2.0 in Shanghai, February 2018.

Source: Author.

(*Xinmin Evening News* 2017). This sentiment was echoed by pro-environmental NGOs in the city:

> Before there were 4 million people not riding bikes, but now they do. From a perspective of urban development, cycling reduces carbon emission, is environment-friendly, and is with many other benefits ... After shared bikes were promoted, ... they have decreased the pressure on public transportation.
>
> (Mr Kuo, Director of Shanghai Bicycle Association, Interview, February 2018)

> This is a win-win solution result with which many benefits can be generated. Sharing bikes is on the one hand promoting low carbon, on the other hand [it has] solved [the] 'last mile' problem, because you cannot extend the subway to your company or your doorsteps. From the viewpoint of NGOs, the business model is easy for us to accommodate, which at the same time results in environmental protection.
>
> (Mr Shen, Shanghai Friends of Nature, Interview, February 2018)

As these comments suggest, PBSS2.0 is valued by stakeholders for its low burden on the city's tax reserves, and its ability to create more seamless low-carbon transport. As one Mobike design manager commented, an important aspect of the design of the system is 'the relationship between government, company and people and how to make the relationship beneficial [...]. So this move is very much like British public–private partnerships and will be the new mood in China' (Mobike Design Manager, 5 April 2017). It is clear that in an era where cities with already over-strained budgets must be seen to be 'doing more with less', many have welcomed these schemes, but in doing so they have enabled potential extraction architectures to be proliferated amongst their citizens with little oversight or regulation.

Cycling, sharing and obfuscation of the unprecedented

One question that arises from this is to ask how this has been able to happen in such an uncontested way. Notwithstanding the fact that state surveillance of citizens is arguably more widespread in China, here we argue that PBSS2.0 has partly eluded critical examination because it has been linked to ideas of cycling and sharing, which share positive associations and obfuscate any less virtuous or transparent motivations.

Cycling has for some time now been synonymous with ideas of liveability and new urbanism: environmentally sustainable, space-efficient and healthy; cycling is seen to be an antidote to the urban ills wrought upon cities by a surfeit of automobility (Bonham 2006; Jennings 2015). Sharing economies likewise have been lauded as more pro-social and pro-environmental ways to organise consumption (Hall and Ince 2017; Richardson 2015). Put the two together in the form of public bike-sharing systems (Stromberg 2017; DeMaio 2009) and surely you have a system beyond reproach and critique; a system whose potential to fix transportation and environmental problems is beyond doubt? Such sentiment is evident in the statements of stakeholders when discussing the emergence of PBSS2.0:

> From the perspective of urban development, PBSS2.0 provides many benefits such as carbon emissions reduction, environmental protection … traffic congestion is quite a serious problem in Shanghai so that shared bikes can replace the traditional transportation such as walking, this can release the pressure of public transportation.
>
> (Mr Kuo, Director of Shanghai Bicycle Association, Interview, February 2018)

> The humble bicycle can not only change people and cities, but can also help change the world – both as a universal symbol of peace, and as a weapon in the fight against climate change.
>
> (Ms Hu Weiwei, in UNEP 2017)

> As bike-share was launched, people thought it was a good idea because it meant green transportation, it was sharing economy, and it appealed to nostalgia … Bike-share was not in a grey area, why the authority would stop that? It contradicts innovation and sharing, doesn't it?
>
> (Mr Chen, Shanghai Municipal Engineering Design Institute, Interview, February 2018)

As we see in these comments, PBSS2.0 is lauded for its ability to reduce emissions, reduce congestion, and increase accessibility. The bicycle is presented as a symbol of peace and sustainability, whilst sharing is positioned as innovative and efficient. As such, PBSS2.0 is connected positively to a lineage of both cycling and sharing, which are positioned as benign and positive solutions to established urban transportation problems. The key point here is that in normalising and rendering the unprecedented benign, it becomes less likely that what may be going on behind the scenes (e.g. data harvesting and sharing) will be problematised. The evidence that this project of obfuscation was successful is exemplified in the comments of one environmental NGO leader:

> Those dock-based shared bikes in the past were in fact a waste of land resource. Now it is very convenient by using GPS positioning (to find a dockless shared bike). The old industrial civilisation that creates pollutants was profit-oriented and capital-oriented. But nowadays human beings can have [a] better relationship with the environment through technology, and we are now reaching the attitude called

> ecological civilisation. We should award a grand prize to the (dockless) bike-sharing.
>
> (Mr Shen Yi-Ke, Shanghai Friends of Nature, Interview, February 2018)

Here we see that PBSS2.0 is positioned as a system outside of circuits of capital, profit and inefficiency in opposition to PBSS1.0 and motorised transport. The apparatus of extraction that forms PBSS2.0 is seen to be benign – a harbinger of 'ecological civilisation' rather than representing a new phase of capital accumulation. This echoes the rhetoric of Mobike founder Hu Weiwei herself:

> As we integrate eco-friendliness, technology, and commerce together, it will turn out to be something extraordinary. We can make eco-friendliness a lifestyle and a sustainable business model at the same time.
>
> (Ms Hu Weiwei, cited in *Tencent News* 2017)

PBSS2.0 therefore can be considered an example of what Zuboff calls the 'unprecedented', meaning that surveillance capitalism is unrecognisably different from what has come before (2019: 12). Zuboff's notion of the unprecedented helps to explain how surveillance capitalism is allowed to proliferate: by using existing lenses and associations to render something that is actually very different as something familiar (and benign), we obscure the original by turning it into an extension of the past (Zuboff 2019: 12). Zuboff gives the example of the term 'horseless carriage' used to describe the automobile in its early days as a way of rendering something strange into something reassuringly familiar. In a similar way, the couching of PBSS2.0 within positive associations of sharing and cycling discourages critical investigation.

That is not to say that this is completely successful, with some interviewees articulating scepticism by suggesting that bike-sharing may not be as benign as it appears:

> Green transportation is the big question mark here. Businesses use the phrase to appeal to people's emotion, to sugar-coat capitalism, and to win public recognition.
>
> (Mr Chen, Engineer at Shanghai Municipal Engineering Design Institute, Interview, February 2018)

As Mr Chen suggests here, PBSS2.0's status as green transportation acts as a concealment for an unprecedented capitalist business model. However, the use of existing categories such as 'green transportation' and 'eco-friendly' fall short in identifying and contesting the facts and inner workings of this new regime. Whilst terms such as sharing suggest an innate pro-social, inclusive and democratic morality to networked forms, Zuboff (2019: 9) argues that we need to look beyond such labels to interrogate these forms, and this begins by understanding and accurately labelling what constitutes them.

PBSS2.0 as apparatus of generation and extraction: scale, scope and action

In this section we argue that contemporary PBSS2.0 is an example of what Zuboff (2019) calls an 'architecture of extraction'. We do this by showing how its rapid growth reflects the need for economies of scale in behavioural surplus; its very presence in fields of everyday mobility suggests it is part of economies of scope that require access to all facets of daily life; and the 'free' provision of the bike/app represents an architecture designed to actively induce action rather than wait for its occurrence. In making these links, we argue that PBSS2.0 is a vehicle for harvesting data as much as (or more than) it is a mode of transportation.

In charting the recent history of surveillance capitalism, Zuboff (2019) shows that the extraction of user data has spread from its confines within early internet architectures into many other aspects of daily life. As she goes on to argue, the ability to create a reliable (and thus valuable) predictive product is reliant on economies of scale, scope and action (Zuboff 2019: 192). We elaborate on each of these in turn below with a particular focus on the ways in which hybridity and movement are essential to the PBSS2.0 extraction architecture and enable the generation and capture of behavioural surplus.

Economies of scale are part of what Zuboff calls the 'prediction imperative' and are necessary because the closer the 'n' is to the entire population, the more accurate the predictions, and the more valuable the product (Zuboff 2019: 192). We see this imperative materialised in the rapid expansion of PBSS2.0. Whilst PBSS2.0 only began operation in Shanghai from spring 2015, the growth of these schemes in the city has been nothing if not meteoric. By February 2017 there were 260,000 public bikes, rising to 630,000 bikes by April 2017, with 7.5 million registered users and ten different operators. By August 2017, there were 1.5 million public bikes in the city – a staggering 16 per resident – with two Chinese firms, Mobike and Ofo, dominating the market alongside many smaller firms. This rapid expansion is paralleled by a rapid global expansion, as these two firms in particular sought to expand the scale of their operations. At the peak of their 2018 operations, Mobike operated in 200 cities in 19 countries and OfO more than 250 cities in 21 countries (Crunchbase n.d.). What this expansion symbolises is an attempt to sign up as many users as possible in as many different territories in order to gain access to behavioural data on a massive scale.

However, scale alone is not enough. As Zuboff points out, 'behavioural surplus must be vast, but it must also be varied' (Zuboff 2019: 193). In moving out from fixed virtual internet architectures into the 'real world',

PBSS2.0 is itself a manifestation of the desire to *extend the scope of extraction* operations. As Zuboff goes on to say, surveillance capitalists understand 'that their future wealth [depends on] ... your bloodstream and your bed, your breakfast conversation, your commute, your run, your refrigerator, your parking space, your living room' (Zuboff 2019: 193). Hence increasing scope actually relies on the (seeming) disappearance and diversification of the internet into mundane objects such as fitness/sleep trackers; voice assistants; doorbells; security cameras; smart watches; mirrors; glasses; air monitors; home and street lighting; healthcare, etc.

As an example of this, the physical apparatus of PBSS2.0 works as an incursion into undefended space. As Zuboff has argued, 'dispossession operations rely on their virtual capabilities to kidnap behavioral surplus from the nonmarket spaces of everyday life where it lives' (Zuboff 2019: 137). This distributed apparatus of smartphone and app, bicycle, data masts, and back-end servers helps to continuously render everyday mobility habits into the digital realm, 'where it finds new life as data ready for transformation into predictions' (Zuboff 2019: 194). Here the act of cycling – of renting a public bike – becomes the means through which behavioural data is generated, and the smartphone/GPS-equipped bike enables data to be captured, transmitted and stored. As Bratton (2009) has argued, the smartphone is not replacing but augmenting and facilitating physical mobility. However, the majority of what the smartphone does is hidden to the user due to the nature of the pervasive computing revolution (Dodge *et al.* 2009; Galloway 2014) – where computers and software become embedded into our everyday lives in such a way that they are rendered invisible.

The provision of the means of mobility in PBSS2.0 also demonstrates a *shift toward economies of action*. Zuboff argues that behavioural surplus must be vast and varied in order to better capture the needs, wants, attitudes, relations, habits and desires of the individual. However, Zuboff goes further to argue that the surest way to predict behaviour is to actively shape it in the form of what she calls economies of action: 'these interventions are designed to enhance certainty by doing things: they nudge, tune, herd, manipulate, and modify behaviour in specific directions' (Zuboff 2019: 193–194). At its most basic level, the provision of bikes to be ridden facilitates a behaviour from which surplus can potentially be extracted. Hence rather than face the uncertainty of waiting for behaviours to occur, PBSS2.0 ensures people will cycle in order to guarantee that behaviours (and surplus) will occur.

The Mobike (and Ofo) app interfaces form a key element of the inducement of users to be mobile. These apps represent three main elements of the journey back to the user as important: time taken – a proxy for transport efficiency – quick workers are workers not stuck in congestion; calories burnt – healthy workers are more productive; and CO_2 saved – the cycling worker is one cutting environmental harm and doing their duty. The app clearly provides a framework for becoming the right kind of subject, and that is to be healthy, green and quick (very similar traits to the subject outlined in Chapter 4). Relating this to economies of action, the app attempts to ensure that users continue/increase the amount of cycling by playing back to them the productivity, health and environmental benefits of doing so. The reasons to do so are to ensure that the apparatus of extraction enables users to meet goals (often set by the apparatus itself) so that they continue to participate in the generation of data. As Galloway has argued, subjects are 'liberated' in the society of control as long as they perform certain forms of mobility (Galloway 2014: 106, in Krivy 2018: 19).

A key point we wish to emphasise here is that these economies of action and associated architectures of extraction are premised upon the increasing convergence and 'always on-ness' of physical and virtual mobility. De Souza e Silva has argued that we are seeing the emergence of hybrid spaces that flow from 'the combination of physical and digital spaces, along with the social use of location-aware technology' (de Souza e Silva 2017: 21). Such hybrid spaces emerge from the conjunction of our ability to be always connected, to be physically mobile whilst in virtual space, and to be virtually mobile in physical space. An interesting side effect of the shift to increasingly hybridised forms of mobility is the ability of such mediums to simultaneously produce, record and combine different elements of ourselves in what has been termed 'datafication' (Millington 2014; Williamson 2015). In its economies of scale, scope and action, PBSS2.0 represents a vehicle for datafication of the everyday social practice of cycling. What is perhaps startling is that such operations have been subject to so little regulation, as we discuss next.

Wither the regulation of PBSS2.0?

According to Zuboff, surveillance capitalists operate in a landscape largely devoid of regulation because they have outrun any attempts to do so and – in a neoliberal landscape that favours as little regulation as possible – caught the few regulators still operating, napping (Zuboff 2019: 10). Whilst China (or any other nation for that matter) does not fit neatly into a neoliberal box, its attitudes towards innovation suggest a neoliberalisation with a focus on deregulation and responsibilisation of the individual. Indeed, interviewees noted that by allowing PBSS2.0 to grow in cities unregulated, the government could avoid accusations that they were holding back innovation and entrepreneurialism:

> Sharing bikes are not only representative of green transportation, but also a sense of sharing and nostalgia. Thus, it is a timely opportunity for the business sector. Government also seems to have to support it, since you have suppressed the 'agitation' left from the regulation of sharing taxis [such as DiDi], and this is a good thing. If you suppress it, this means you have violated the national trend of innovation and sharing.
> (Mr Chen, Shanghai Municipal Engineering Design Institute, Interview, February 2018)

As this suggests, after demanding that app-based car-sharing firms Uber and DiDi must be regulated to operate, the Chinese government has been keen to take a more 'watch and wait' approach to PBSS2.0. This is in no small part because it does not want to be seen as stifling innovation and entrepreneurialism. However, with well-publicised scenes of dockless bikes piling up in the streets (see Figure 7.3), by August 2017 the Shanghai municipal government called a halt to any new bikes appearing on the streets and drafted a set of regulations for the operation of PBSS2.0. The draft of 'Regulations on the Internet Bike Rental Management of Shanghai Municipality' gave city districts powers to ask firms to reduce the numbers of bikes, manage their parking, or cease operations altogether (Li-Ning Ouyang, cited in *The Paper* 2018) (see Figure 7.4). In response, Ofo stated that it would:

> cooperate with the local government to develop a parking demonstration zone and initiate regulations for parking and car management. Furthermore, Ofo also seeks to integrate BDS [Baidu Navigation System] to upgrade electronic fence system so as to develop virtual parking areas.
>
> (Lu Chen, cited in *China Youth Daily* 2017)

What is salient here is that regulation applied only to the physical – and all too visible – aspect of the apparatus; the invisible back-end gathering user data, as with other forms of surveillance capitalism, has been left to operate unchecked and unregulated.

PBSS2.0 investment and the Chinese internet giants

Having established some operational characteristics that place PBSS2.0 in the same frame as other modes of behavioural surplus extraction, in the following section we highlight the investment patterns that have driven this expansion. In doing so we argue that this pattern echoes those of established surveillance capital firms like Google and Facebook described by Zuboff (2019). However, whilst these investments (and the comments of company employees) predict a turn toward surveillance capitalism as the dominant source of profit for PBSS2.0, we want to start here by grounding the origins of Mobike and Ofo in pro-social goals, where management characterise them as companies driven by a desire to solve urban problems and improve the lives of citizens:

> Shared bikes have already transformed people's mode of transportation. Mobike data shows the total biking distance of its users in China has surpassed 2.5 billion kilometers since April 2017 … It is said to have put the last piece on the public transit map.
>
> (Dafei 2017)

Figure 7.3 Excess dockless public bikes piling up, February 2018.
Source: Author.

Figure 7.4 Mobike operative tidying dockless bikes outside a Metro station.

Source: Author.

> Bikesharing has changed Chinese urban residents' mobility habits – people take 55 percent fewer trips by cars, and illegal motorcycles that used to provide 'last-mile' transportation solutions have been reduced by 53 percent. Every shared bike is a valuable data source that can help us to better understand urban residents' behaviour and to plot a zero-carbon path for not only Chinese cities, but cities around the world as they begin to adopt this technology.
>
> (Tan and Defei 2018)

As these statements attest, whilst they are profit-driven companies, much like the leading surveillance capitalist firms, PBSS2.0 operators appear to be driven – at least initially – by a desire to make the lives of citizens and the management of cities more efficient. This is one of the first similarities between PBSS2.0 firms and more established surveillance capital firms. According to Zuboff (2019), Google started off with a pro-social mission that led it to feed all of its behavioural surplus (information on user behaviour gleaned from their online activity) back into improved search functions for users: in its early days it did not monetise this information. In line with this, Mobike and Ofo have fed much of their initial energies into enhancing the design/availability of bikes and quality of the service.

However, due to the restless demands of the venture capital firms investing in its early rounds of funding, Zuboff goes on to state that Google was forced to focus on harnessing and monetising user behavioural data to support and improve its advertising arm. It was Google's focus and consolidation of this that saw the company's profits rocket in the early 2000s (Zuboff 2019). A similar trajectory can be seen in the PBSS2.0 industry, with massive investment from venture capital and internet search giants in the hope that PBSS2.0 will become profitable.

PBSS2.0 inward investment

When looking at the financial performance of Mobike and Ofo, it is clear that bike rental in and of itself is not a sustainable business model. The financial statements relating to Meituan-Dianping's acquisition of Mobike in 2019 showed that Mobike's monthly deficit was up to CN¥500–600 million (£50–60 million). Ofo, at a similar enterprise scale with Mobike, also spent around CN¥ 600 million on monthly operating costs. To give another example, the fast-rising Hellobike (another bike-share firm in China) had a reported 2017 revenue of CN¥128 million (£13 million), yet its net loss for the year was four times greater at CN¥480 million (£50 million). Such numbers suggest that renting bikes alone is not anywhere close to putting these companies into profitability. It follows that another source of revenue will have to be found soon if they are to survive.

Our interviews with industry representatives confirmed that the data generated by usage of the scheme could be a valuable asset to the companies even if they were not yet sure how to analyse and monetise it: 'the big data is very valuable. We have the name, bank account, ID, workplace, address; … it is a goldmine' (Mobike Government Relations Manager, 5 April 2017). However, it was recognised that the data generated by PBSS2.0 usage could be more valuable when combined with other data that was facilitated at Mobike through integration with other apps, as the Mobike Design Manager noted:

> just yesterday … Mobike has become a small function of Wechat so maybe this data can communicate with other data. But for the Mobike data there is no profit unless you can bring it together with other data, then perhaps you can make money because this data combines together. So in every scenario we try to combine to have more possibilities.
>
> (Mobike Design Manager, 5 April 2017)

The value of combining and sharing user data gathered through web-based apps is not new. Frith (2013), for example, has discussed the ways in which location-based social network apps encourage users to sign up across platforms in order to combine data on users and gain a more holistic understanding of user preferences for advertising purposes. Similarly, Ossandon notes how data can be made more useful 'by *combining* individual trackers. For each is involved in a quite distinct form of "socio-technical knitting"' (Ossandon 2013, in Deville and Van der Velden 2016: 99). This same drive is evident in the comments above, where the ability to link datasets together is seen as vital to the future profitability of Mobike's operations. The emphasis on data, where it is being sent, and how it is being combined, strongly suggests in line with Zuboff (2019) that for PBSS2.0 operators, 'serving the genuine needs of people is less lucrative, and therefore less important, than selling predictions of their behavior' (Zuboff 2019: 92). In such a reading, the mobility of the PBSS2.0 user becomes the object from which raw materials are extracted in order to be processed into prediction products.

The post-2008 financial crisis has proved a fertile ground for the expansion of surveillance capitalism into new territories. Zuboff suggests that it was the period directly after the crash that 'set the stage upon which surveillance capitalism flourished' (2019: 492), with many investors licking their wounds and looking for somewhere other than the property market to get a good return on their investments. The huge profits generated by the likes of Google and Facebook have signalled to investors that there is money to be made by investing in companies that could tap into and monetise behaviour. Between 2000 and 2004, for example, Google's 'discovery of behavioral surplus had produced a stunning 3,590 percent increase in revenue' (Zuboff 2019: 87). Zuboff (2019) argues that:

> the already visible trend … is that any actor with an interest in purchasing probabilistic information about our behaviour and/or

> influencing future behaviour can pay to play in markets where the behavioural fortunes of individuals, groups, bodies, and things are told and sold.
>
> (Zuboff 2019: 96)

The bike/phone hybrids deployed by companies like Mobike and Ofo represent what are potentially lucrative architectures of extraction for would-be investors. Investigating who has funded the expansion of the two largest PBSS2.0 companies since 2016 suggests that investors believe that despite a demonstrable inability to make money from bike rental, these companies may still produce monetisable products.

Mobike is headquartered in Beijing, China and was founded 27 January 2015. Mobike's early funding rounds were dominated by small venture capitalists specialising in internet and disruptive technologies and characterised by modest investments. Later funding rounds in September 2016 brought in more established venture capital firms such as Sequoia and Hill House as well as Chinese internet giant Tencent. However, it was only in funding rounds D and E from early 2017 that the money really rolled in, with the likes of Tencent putting in the bulk of US$600 million. By June 2017 Mobike had secured funding of almost US$1 billion and was valued at US$3 billion in the same month (Crunchbase n.d.). Similarly to Granovetter and Ferray's findings in their study of Silicon Valley venture firms, 'a connection with a high status VC firm signals the high status of the start-up and encourages other agents to link to it' (in Zuboff 2019: 73). Despite a downturn in the outlook for PBSS2.0 in 2018, Mobike was acquired by Chinese web company Meituan-Dianping for US$2.7 billion in April 2018. A name change from Mobike to Meituan Bike was confirmed on 23 January 2019 (Crunchbase n.d.). This was followed shortly after by the appointment of Liu Yu as president of Mobike. Liu was previously a special advisor to the company and (pertinently) the general manager for internet giant Alibaba's language service platform (Borak 2018).

Ofo, the leading competitor to Mobike, is also headquartered in Beijing, China and was founded in April 2014. Its funding trajectory has been similar to that of Mobike, with early rounds dominated by small venture capital and later rounds E and F in July 2018 and March 2019, respectively, dominated by Asian internet giant Alibaba. To date it is has raised US$2.2 billion in funding, almost US$900 million of which was raised in round F alone (Crunchbase n.d.). Recently, however, it has been fined for defaulting on debt in December 2018, sued by suppliers, had assets frozen (Li 2019) and has been banned from operating in Singapore, one of its main territories (Dowling 2019).

Given that bike rental in itself does not currently pay the bills, the investment patterns of these two companies tell us a lot about what it is hoped the sources of revenue might be, and how their focus might shift if they are to survive. There are two key clues as to which direction Mobike and Ofo might take. The first is that the fortunes and funding of PBSS2.0 companies are following the same trajectory as those of companies like Google in that they are being driven in part by venture capital firms. Just as Google's focus on monetising user search data in the early 2000s was driven by the restlessness of investors to see a return on investments, it is very likely that PBSS2.0 companies are being subject to similar pressures, with investors likely to look elsewhere if a source of profitability cannot be found.

The second clue is given by the presence of the internet giants Tencent, Baidu and Alibaba amongst the main investors of PBSS2.0. Zuboff asks the question, 'If Google is a search company, why is it investing in smart-home devices, wearables, and self-driving cars? If Facebook is a social network, why is it developing drones and augmented reality?' (Zuboff 2019: 127). We can ask a similar question in relation to PBSS2.0: if Tencent and Alibaba are search companies, why are they interested in bicycle rental? The most convincing answer is that these two – Tencent and Alibaba – companies, which have a pre-existing interest and expertise in extracting and analysing behavioural surplus data, see in PBSS2.0 a way of expanding the scope of their extraction operations by gaining access to data regarding another social practice, in this case daily mobility. As Zuboff goes on, an interest in a wide array of activities that appear to be 'scattershot across a random selection of industries and projects are actually all the same activity guided by the same aim: behavioral surplus capture' (Zuboff 2019: 127).

The interest of Tencent and Alibaba in the seemingly unconnected world of bike-sharing echoes the investment patterns of Facebook, which, for example, has in the last seven years purchased a wide range of companies including messaging firm Instagram (2012); gaming developer Oculus VR (2014); drone manufacturer Ascenta (2014); and health monitoring firm ProtoGeo Oy (2014). Similarly, given Google's varied acquisitions, its:

> stores of behavioral surplus now embrace everything in the online milieu: searches, e-mails, texts, photos, songs, messages, videos, locations, communication patterns, attitudes, preferences, interests, faces, emotions, illnesses, social networks, purchases, and so on. A new continent of behavioral surplus is spun each moment from the many virtual threads of our everyday lives as they collide with Google, Facebook, and, more generally, every aspect of the internet's computer mediated architecture.
>
> (Zuboff 2019: 127)

In line with this, the investment in Mobike and Ofo by the leading Asian internet providers suggests they have similar goals of gaining access to an ever-broader range of user data.

Extracting data

It is worth pausing for a moment and asking how we can be sure that data is being extracted and used for profit. After all, when interviewed in the media, bike-share companies have been at pains to point out that they don't sell user data, as Bliss has reported:

> Mobike does not share or disclose any personal data to any third party without the users' consent or in any way that is non-compliant with local data protection laws. Mobike prioritizes user privacy, and any data collected is anonymized before sharing with local, U.S. cities.
>
> (Mobike representative, quoted in Bliss 2018)

Other representatives have reiterated this position, with another Mobike representative stating that, 'we don't make money through selling data' (Nada, quoted in Teffer 2018). This position was echoed by representatives from both Ofo and O Bike, stating that they don't sell user data and were focused on bike trips in order to make money (Teffer 2018).

Moreover, many industry commentators have been sceptical that the data generated through PBSS2.0 could be worth anything, with one commercial director noting that, 'unless data is highly differentiated, scalable and useful to other businesses or sectors, it's got questionable spectrum appeal and therefore value' (Claire Lawler, Director of Product Commercialisation at Fairfax Media, quoted in Teffer 2018). However, there are at least three reasons why such scepticism seems misplaced.

First, with queries about whether user data will be sold, people are asking the wrong question: given that the two big PBSS2.0 players left in the market are both now owned by internet companies, as outlined earlier, there is no need for data to be 'sold' to these companies in order for it to be analysed and used. As Bliss reports, Mobike's privacy policy reads: 'by signing up, accessing and/or using the Services, you expressly consent to our transmission, processing and storage of your information in locations outside the United States or your country or region of residence' (Bliss 2018). Hence there is no need to 'sell' the data to companies interested in manufacturing predictive products; ownership ensures that the raw materials produced by the subsidiary are available for incorporation in production processes.

A second related point is that the data itself is not the product here; the data is the raw material, so it is unlikely that these companies would want to sell the raw material – as Lawler states (quoted in Teffer 2018), it is unlikely to be worth much. Rather, they want to sell the finished predictive product with the value added in order to realise a profit. So again there is a semantic issue at stake in that not only is data not being sold, if anything is sold, it will be the (predictive) product and not the raw material (data). So the right question to ask of PBSS firms is whether the products derived from user data will be sold?

A third point is that despite terms and conditions stating that data will not be sold or shared, it remains a common occurrence. Vallina-Rodriguez *et al.*'s (2016) study of app ecosystems, for example, found that despite contractual obligations, sharing of data with third parties and affiliates is the norm (1). This calls into question whether companies abide by or can be effectively policed with regard to the sharing and sale of personal data.

Following the data

Even if we agree that PBSS2.0 companies wish to monetise data, can we verify that the data from PBSS2.0 usage is being transmitted to the parent companies? Whilst on the face of it this is difficult given the secrecy around the practices of surveillance capitalists, Deville and Van der Velden (2016) help to point the way. Susan Leigh Star (1991) asks a pertinent question with regard to how to make the invisible, visible in social enquiry. In line with Zuboff, she argues that forms of social control are tied to the erasure of practical activities that stabilise them. Certainly, Vallina-Rodriguez *et al.* (2016) suggest that researchers and regulators have very little idea of what goes on within third-party internet ecosystems.

Taking up this challenge with regard to machine labour, Deville and Van der Velden (2016) suggest a focus on 'tracker tracking' in order to uncover the 'automated, unseen, digital work undertaken by online data gathering tools, many provided by third party providers, activated when a user visits a particular website' (90). Deville and Van der Velden go on to argue that following the work of trackers (such as web browsing cookies) built into websites and apps is a key way we can find out about how 'companies engage in online profiling' and 'weave data together into more or less personal profiles' (Deville and Van der Velden 2016: 101).

Using this method via the Lumen tracker for Android phones (available from the project Haystack website) in relation to PBSS2.0 is instructive. Coded by researchers at MIT, the Lumen app 'helps mobile users understand how their mobile apps handle their private information, including the sensitive data their mobile apps leak and with whom they share it' (Vallina-Rodriguez *et al.* 2016: 2). Using the app in conjunction with the apps of Mobike and Ofo, it becomes clear that usage data is indeed being sent to the parent companies. Using the Mobike app, for example, shows that it sends a range of user and usage data back to Tencent (field diary, June 2019). Our own experiments to find out where data is being sent echo the large-scale findings of Vallina-Rodriguez *et al.*, who studied around 3,000 apps and found that amongst the sites that user data was most commonly transmitted back to were Facebook.com and Googleadsmanager.com (2016: 2). Whilst far from conclusive, it appears that PBSS2.0 user data is being sent to the respective big internet investors. What is less clear is exactly what is being sent and what it might be used for. Having established the presence of an architectural and investment context that positions

PBSS2.0 as a likely source of raw material for predictive products, in the next section we discuss some of the possible products that could be manufactured from this data.

PBSS user data and the manufacturing of predictive products

When the Chinese website Meituan-Dianping purchased Mobike in 2019 it assumed roughly $700 million in debt (Teffer 2018). Certainly, as Teffer (2018) has argued, with high rebalancing costs, low hire fees and high rates of vandalism and attrition, it is difficult to see how PBSS2.0 can become profitable. Accordingly, we argue here that in line with its more established surveillance capitalist forebears, the future success of PBSS2.0 lies in its ability to create predictive products out of the data it gathers. But if this is the case, what kinds of products might these be? In this section we highlight the ways in which PBSS2.0 app interfaces and corresponding points system (as seen in the Mobike app) not only regulate and encourage use, but also gather valuable data that can be used to create predictive products to be sold on behavioural futures markets. We discuss three examples of predictive transport models; route data used to create advertising opportunities to retailers; and app data indicating trustworthiness as three potential revenue streams.

Transport modelling data as predictive product

One of the most obvious ways that PBSS2.0 companies can monetise data is through using users' trip patterns to understand gaps in the public transport network and where the city might build or improve cycling facilities. An article co-authored by Mobike's chief data scientist states that, 'Mobike is partnering with seven Chinese cities [...] to use their data for evidence-based urban planning in public transit gaps, property development density, and to facilitate intelligent transport dispatch' (Tan and Defei 2018).

One of the reasons why these kinds of real-time user data can be so valuable in such contexts is because of their actuation capabilities:

> the real power is that now you can *modify* real-time actions in the real world. Connected smart sensors can register and analyze any kind of behavior and then actually figure out how to change it. Real-time analytics translate into real-time action
>
> (Zuboff 2019: 278)

Indeed, Defei (Mobike chief engineer) goes on to state that PBSS2.0 data is a much more granular, 'real-time expression of each city's density and the transportation dynamics between where people live and where they work', and that alongside new open-source standards, it should be possible to aggregate and analyse data from different companies (ibid.). Such opportunities were also something talked about by planners and industry representatives in our interviews:

> It is valuable to use big data for city's future planning such as old town regeneration. Nevertheless, it is also very helpful for management since the bicycle is mobile and the index completely based on real situations.
>
> (Mr Wang, manager of FOREVER Company, Interview, February 2018)

> From the government's side, big data is very good to analyse and utilise, but has yet to start ... we consider to gather data from parking, motor vehicles and non-motor vehicles use ... we are now working on parking problems and we will ask them [PBSS2.0 companies] to give us data for analysing.
>
> (Mrs Wu, Director of Shanghai Pudong New Area Construction and Transportation Commission, Interview, February 2018)

As these comments suggest, interviewees from different Shanghai municipal departments confirmed that they had an interest in using the data from PBSS operators to assist analysis of parking and route requirements of non-motorised vehicles. This confirms that there is a potential market for real-time predictive products manufactured from PBSS2.0 user data. However, it was also evident that PBSS2.0 operators were still trialling the analytic-industrial processes through which they could add value to the data. Interviewees confirmed that the big PBSS operators had collaborated with university research institutes to understand how to analyse the data they had, but that this had not yet made its ways to government departments with operational responsibility:

> The PBSS enterprises usually find and provide funding to research institutes, and institutes conduct data analysis and report for them [...] This research usually refers more to symbolic functions or business lobbying; it is like Google develops a big data system for information development and management, in order to distinguish itself from other companies: I am Mobike and you are not Mobike; I have big data and rational analysis, and you don't.
>
> (Mr Shi, Director of Shanghai Urban Planning and Design Research Institute, Interview, September 2017)

> Mobike has set up a public digital platform with THUPDI (Tsinhua Planning Design Institute) and Ofo has cooperated with the Research Institute of Highway Ministry of Transport. I think the quality of data that the company gives research institutes is poor because these are

> commercial secrets ... and why would they share the real situation with the Institute or the public?
>
> (Mr Chen, Shanghai Municipal Engineering Design Institute, Interview, February 2018)

As these accounts suggest, there is some scepticism that a way will be found to turn the raw data materials into predictive products, but there is clear evidence that production processes are being trialled. There are two further aspects of these statements we want to draw attention to: the first is the sense that far from being a public good to be open-sourced, user data is closely guarded. Second, the limited data-sharing occurring seems to be more about presenting PBSS2.0 operators as part of the 'new China' as home to innovative high-tech data analytic companies rather than companies dealing with 'old technology' such as bicycles. Here data becomes part of a broader branding exercise. Corroborating this, it was evident from our interviews with operators that rather than share location-based user data with the municipality so that they could know where to target improvements for cycling, PBSS2.0 operators wanted to use ownership of this data to increase their brand profile and shape the conduct of the government:

> We will figure out which parts of the data are valuable [and utilise it] in resource exchanging. So, for example, if the mayor wants to say something good about Mobike or we have a district encouraging Mobike colours then we might give more data to them. But if a government tries to force the data out of us then we are probably going to say no.
>
> (Mobike Government Relations Manager, Interview, 5 April 2017)

What we draw attention to here is that the creation of the mobile hybrid and subsequent ownership of user data enables the PBSS2.0 operator to leverage the government because of both its use value to planning officials and district offices, and de facto privatisation. With PBSS1.0 – much like in other cities such as London and Paris – user data is generally owned by the city government because they subsidise and govern the service, even if its day-to-day operation is subcontracted. In PBSS2.0 this is no longer the case and the data may be used as a bargaining chip to increase the brand image of the operator, or certainly have a price attached to it. Of key interest here is the way in which the creation of a predictive product serves to shape the conduct of the government in order to produce a preferential business climate. The point we emphasise is that rather than a shared partnership between public and private, there is evidence here that this relationship is becoming *more* unequal, with any goal of maximising collective utility becoming subordinated to maximising private utility. Indeed, we would argue that the ability to own citizen data and create predictive products is producing shifts in urban transport governance.

A further point to highlight here is that through this process of creating predictive products from citizen data, 'urban citizens become sensing nodes … or citizen sensors … a modality of citizenship that emerges through interaction with computational sensing technologies' (Krivy 2018: 15–16). Here citizens become 'empowered' as an 'aggregated dataset' (Birchall 2016: 7) but that empowerment does not extend to participation in acting upon the data (Krivy 2018: 16). As Gabrys goes on, whilst we might aspire for planning policy to be guided by democratic principles of equality and justice, 'the very responsiveness that enables citizens to gather data does not extend to enabling them to meaningfully act upon the data gathered, since this would require changing the urban "system" in which they have become effective operators' (Gabrys 2014: 43, quoted in Krivy 2018: 16). The outcome is what Birchall calls 'shareveillance' – 'a state in which we are always already sharing; indeed, in which any relationship with data is only made possible through a conditional idea of sharing – produces an anti-politicised public caught between different data practices' (Birchall 2016: 1–2). For Birchall (2016) the issue at stake is less that of incursion into personal practice 'and more the anti-political act of only imagining the public as an aggregated dataset' (7).

The use of mobility patterns in predictive products for commerce

Existing literature on surveillance capitalism also gives insights into other potential ways in which PBSS2.0 user data can be utilised. To be clear, there is currently no firm evidence that user data is being used by PBSS2.0 operators in the ways we suggest are possible here, only that other movement-based extraction architectures have utilised data in such a way, and that industry representatives have expressed hopes that their data can be used for similar ends:

> Ofo's whole business strategy … [includes] … B2B (business-to-business), finance, local lives, and business blockchains by integration among bike advertising, web advertising, and business annual membership.
>
> (Dai Wei, Ofo CEO, cited in Chong 2018)

> The higher value of bike-sharing lies in the cycling data … bike-sharing enterprises may develop many interesting products based on the cycling data, which can be beneficial to the bike-sharing industry itself and may further export to other fields.
>
> (Dang Xiong, cited in Chong 2018)

The use of route data that Dai Wei refers to above, and its use to target advertising and charge retailers for 'sponsored locations', is one of the more obvious ways in which data can be utilised. Mobile marketers encourage 'life pattern marketing' to understand daily behavioural patterns to better predict future behaviour and intercept aspects of daily routines with targeted messaging (Zuboff 2019: 231). The key precedents here are location-based gaming apps like Pokemon Go. Only six days after the release of Pokemon Go, reporter Joseph Bernstein advised users to check the amount of data being collected, with the game app recording 'where you go, when you went there, how you got there, how long you stayed, and who else was there' (Bernstein, quoted in Zuboff 2019: 299). The game developer Niantic can use this information to both target advertising to these locations and times, but more importantly, for a fee from retailers and restaurants, it can build particular locations into the game to encourage users to go there. As Zuboff has stated, 'the game had demonstrated that it was possible to achieve economies of action on a global scale while simultaneously directing specific individual actions toward precise local market opportunities' (Zuboff 2019: 299).

A similar scenario can be envisaged with PBSS2.0 because the operator will know which routes users take and will be able to target advertising accordingly, particularly when combined with other data regarding income, age, etc. This became more of a reality for Ofo when it announced its intention to build Baidu maps into its interface (Lu Chen, cited in *China Youth Daily* 2017). As Huang (2018) notes in his commentary, the acquisition of user geo-location data is a key driver of Tencent and Alibaba's investment in PBSS2.0. Indeed, Zuboff argues that the easiest way for surveillance capitalists to render reality into behavioural surplus is 'to know exactly where you are all the time' (Zuboff 2019: 231). She goes on to say that the reason most smartphone apps demand to access location even when not required is because it is lucrative: 'retailers use "geofencing" to demarcate a geographical area and send alerts to smartphones within those parameters: "Come here now!" "Buy this here!" "An offer, just for you!"' (Zuboff 2019: 231). Following on from this – and similarly to mobile gaming – the ability to decide where users should park bicycles via geofencing means that PBSS2.0 operators could guarantee higher footfall in particular areas and therefore charge retailers and restaurants to advertise in these locations. In such a scenario the mobility of users becomes a key source of data for predictive products that are attractive to advertisers for their promises of more accurate targeting.

Mobility and trust

A third possible revenue stream is to use PBSS2.0 users' mobility patterns and behaviours to profile them in similar ways that age, income, gender and ethnicity are currently used to profile and predict behaviour (Deville and Van der Velden 2016: 93–94). The precedents for this come largely from the insurance and financial services industries (e.g. credit scoring). Studies by IBM related to Twitter accounts have shown that a small number of measureable personal characteristics, such as 'the need for love', 'openness' and 'neuroticism', can be measured from social media profiles and used to predict whether people will like a brand and whether they will click through on particular advertisements: 'people whom the machines rated as moral, trusting, friendly, extroverted, and agreeable tended to respond, compared to low response rates from people rated as cautious and anxious' (Zuboff 2019: 263). Evidently the credit rating function of the Mobike app has the capacity to render similar user data.

Through Mobike's big data platform 'Magic Cube', Mobike records a user's cycling history, payment records, how and where they park the bike, how fast they ride, whether they return it on time, and combine these with other user information on the platform (such as age, address, job, etc.). Based on this information, Magic Cube evaluates and calculates a user's 'credit' level. The better 'behaved' a user is, the higher the rating. Currently a good credit rating (including parking bikes well, returning others' bikes) can generate 'Mobike coin' to exchange for products and services. But beyond this, such data can be used to make an assessment of characteristics such as trustworthiness, which as the IBM example of Twitter users has demonstrated, can be a valuable predictor of user behaviour in other realms of social life. The Mobike app and its credit mechanism therefore can gather the raw materials required to manufacture a valuable predictive product of interest to insurers and advertisers.

Such data, however, could also be valuable to other financial services such as those involved in credit scoring and personal loan provision. As Deville and Van der Velden (2016) have argued, big data scoring combines data to assemble a datafied user to create 'very rough proxies for an individual' (93–94). Mobility patterns are likely to be a useful additional resource in this because they say something about how you travel, the different shops and places you visit, with what frequency, in addition to how 'well' behaved you are with the bike. As one of the leaders in the field of credit data analytics has stated, 'all data is credit data' (Zestfinance, quoted in Deville and Van der Velden 2016: 93–94). Thus the behavioural 'exhaust' of users' mobility practices has the potential to be repurposed as the raw material for predictive products.

Importantly these products not only reductively represent the user but can also produce the user when they are utilised by companies and begin to enable and constrain the consumption choices of that consumer, and start to shape the 'real' user in the form of the datafied proxy user. For example, whether such data may start to dictate whether that user will be accepted for a loan and which advertisements are targeted at them. As Smith argues:

> data profiles ascribed to an observed body progressively structure how its occupant experiences social life. How a body is encoded by a surveillance system (i.e. as desirable or as risky) has a significant impact on the treatment a person will receive in certain situations.
>
> (Smith 2016: 109–110)

Hence through PBSS2.0 it may be that mobility joins the ranks of particular reductive indicators like race, ethnicity, class, age, gender and sexuality used in practices of data profiling, and which reflect and reify cultural biases and perpetuate social inequalities (Smith 2016: 109–110).

Absence of institutionalisation and the uncertainty of success

As the previous discussion has demonstrated, much of the ability for PBSS2.0 to become financially self-sustaining remains speculative: whilst PBSS evidently gathers substantial raw materials, it has yet to refine its production processes to the point where it has any viable products. Added to this, its economies of scale have been reduced since their 2017 peaks and it has withdrawn in most cities as rapidly as it deployed. As Zuboff (2019) has argued, the success of any new economic mutation is far from certain: 'occasionally, the elements of a mutation find the right environment in which to be "selected" for propagation … but it's even more likely that potential mutations meet their fate in "transition failure"' (Zuboff 2019: 47). Certainly if a mutation is to be reliably sustained, its new aims and practices must be translated into new institutional forms (Zuboff 2019: 51). Çalışkan and Callon (2009) in particular are unequivocal on the importance of institutions in processes of economisation because of the role they play in enhancing or bringing into existence the competencies that may exist in human beings, thus enabling the 'economic formatting of individual behaviours' (Çalışkan and Callon 2009: 380). Institutions in this reading are responsible for presenting and framing practices in particular ways and an absence of institutionalisation can mean that practices are not successfully economised. Here we argue that an absence of institutionalisation within PBSS2.0 has manifested in three ways.

First, PBSS2.0 has been rolled out in many cities where cycling as a social practice is weakly institutionalised: in most towns and cities across the globe it lacks the material environments (e.g. separated infrastructure), positive socio-legal frameworks, and cultural traditions that would enable it to flourish. Many of the cities it has been implemented in have very low mode share of cycling (around 2%), meaning that it is unable to attract new users and achieve real economies of scale or scope. As such, a core component of the required extraction architectures is missing/inadequate and beyond the power of the surveillance capitalist to control.

This weak institutionalisation of cycling as a form of mobility is compounded by a weak institutionalisation of sharing as a concept. Whilst PBSS2.0 is more accurately a form of rental, given its embeddedness in corporate structures, its successful use still requires aspects of reciprocity associated with shared resources. In neither Eastern nor Western societies is there a significant culture of sharing. Bardhi and Eckhardt (2012) in their study of user attitudes to car-sharing firm Zipcar noted that users did not feel comfortable with the cars, created no positive associations and ultimately treated the cars with contempt, damaging them and leaving them unclean for subsequent users (889–891). Overall their study suggested that the ultimate goal of users in utilising the shared resource was the maximisation of personal utility, with ownership still a more desirable end state than sharing. Similar issues have been well publicised in relation to PBSS2.0, with high rates of equipment attrition and bike abandonment negatively affecting profitability. The scale of this problem has led to the highly publicised collapse of some PBSS2.0 operators, such as WuKong bike, which had 90% of its fleet lost, stolen or damaged within its first five months of operation (Everington 2017). Whilst none of the operators we talked to had problems of this magnitutde, the scale of bike loss was deemed significant by Ofo, though not precisely known:

> It's hard to know the [damage rate] since the number of bikes updates constantly. It is difficult to count and I guess people inside the firm do not know that either. In fact, the management of bikes cannot align with the rapid expansion of this industry.
>
> (Mr Xing, former Regional Manager of Ofo, Interview, 6 December 2017)

Mobike, on the other hand, were able to put a figure on their losses, with one manager estimating that they had at any one time around 30% of their fleet going missing, and one crew member commenting that 'there are definitely 200–300 malfunctioned bicycles checked and fixed a day, and approximately 5,000 bikes a day in the whole city returned to warehouse for maintenance' (Mr Chen, cited in Chong 2018).

On a conceptual level, the abandoning of bikes anywhere on the streets is emblematic of the maximisation of private utility (saving time and effort) over collective utility (the ability of other users to easily use the public realm) and suggests that concepts of sharing are too weakly institutionalised in modern societies for such schemes to be successful.

The final point relates to the fact that surveillance capitalists only have a limited time to streamline and institutionalise their production process and turn their data into profitable products before investors begin looking for other places to invest their money. As Reid has noted:

> Investors aren't interested in the bikes in dockless bike share systems and they only want to mine the data. It's possible that tech companies will be looking for a quick return on the investment and will possibly go on to feast on another data-rich sector some time soon, leaving an awful lot of share bikes to rot by the roadside.
>
> (*Bikebiz* 2017a)

Huang echoes this view, stating that:

> the mainstream view – pushed heavily by investors – was that the user data, rather than the service, was the actual product, and that the

> harvested user data would lead to more targeted advertising. But as so often occurs, these ambiguously defined long-term returns yielded nothing in the endless interim.
>
> (Huang 2018)

As a result, the winners in this particular surveillance sector may well be those who can afford to stay in the market the longest, and this favours those with the greatest investment who can remain unprofitable for the longest time. Investment by Tencent, Alibaba, and Meituan-Dianping, all of whom have deep pockets, are the only reason Mobike and Ofo remain afloat. However, given the accumulated losses, it is likely that if the processes required to create profitable predictive products are not institutionalised very soon, these investors will also cut their losses and withdraw.

Conclusions

The primary purpose of this chapter has been to demonstrate that PBSS2.0 companies are surveillance capitalists, or at least very likely in the process of becoming so. In contrast to a logic of the unprecedented that positions them as benign transport services, we have shown through a variety of empirical and circumstantial evidence that the most convincing interpretation of what PBSS2.0 represents is a manifestation of surveillance capitalism due to its origins, investment, ownership, practices, and potential predictive products.

Having demonstrated that PBSS2.0 as an extraction architecture is geared towards the scraping of behavioural surplus, we argue that this positions PBSS2.0 as another kind of 'mobility fix': one geared towards harnessing the labour of the cyclist to produce raw materials that can be manufactured into predictive products. We emphasise the idea that fixes are increasingly found not only in the financial realm (temporal) or the spatial, but in the social realm, and in particular in the realm of mobility through the conjoining of virtual and physical to create hybrid mobiles that enable the economisation of social practice. This focuses our attention on the relations between physical movement between places, and the relationship to a reduced, attenuated and modified user based upon that movement. Through this process, particular aspects of daily mobility are foregrounded as profitable, representing a shift in the ontological status of cycling as particular aspects (that can be measured, analysed and monetised) become more or less important. Hence, what is potentially lost in the drive to a supposedly pragmatic, neutral and common-sense urbanism (Kitchin *et al.* 2018: 22) are the 'social, cultural, historical, institutional and political complexities' (ibid.) of daily mobility that give it its unique character. In its conjoining of virtual and physical mobility, PBSS2.0 represents a new frontier in the drive to economise previously 'unproductive aspects of mobility' into the economic realm. Once again, rather than being materialised as a force enabling us to question logics of unrestrained growth, cycling is being brought into circuits of capital in a way that reproduces or accelerates accumulation in unsustainable ways. There are a number of implications from this that we wish to draw out.

The first is that in line with examples in other chapters, the sustainability of contemporary manifestations of cycling must be questioned. If the constant search for new markets and maximising economic growth is the key challenge to be overcome in the anthropocene, as Jackson (2009) has argued, utilising cycling as a vehicle to extract surplus to sustain growth positions it as central to accumulation, rather than as a vehicle through which such logics could be questioned. However, the technologies that comprise PBSS2.0 could be configured in very different ways if underpinned by different logics. As Zuboff has argued, 'technology in modern times comes already patterned with an economic orientation' (Weber 1978, in Zuboff 2019: 16). In line with this, PBSS2.0 represents the expression of a specific logic of accumulation, though clearly one that is contingent and not fully strategised, as we have shown.

This absence of sustainability is not just related to a logic of continued growth; rather it is also related to a lack of transparency, citizen participation, and uneven governance in defining what data is gathered, where it goes and how it is used. The mode of operation of PBSS2.0 has profound implications for citizens given the opacity of mobile data harvesting infrastructures (Vallina-Rodriguez *et al.* 2016). As Birchall (2016) has argued, the key issues here are political: whether citizens have any say in analysing their data, or how predictive products might 'act back upon' and shape the possibilities for individuals (Smith 2016). As Birchall goes on, the fact that citizens only count in terms of their role as flat data affects the scope of political agency (Birchall 2016: 7). The point we draw out here is that, much like imagining cycling as only useful in a narrow form, such as commuting, marginalises alternative forms of agency (as in Chapter 4); seeing PBSS users as sources of behavioural surplus further flattens political agency because the harvesting and use of it is opaque and exclusive to users.

Moreover, the emergence of PBSS, and by extension all forms of mobility as a service (MaaS), has implications for urban governance as it relates to transport. The proliferation of private providers in these markets means that data on how the city functions is increasingly a private resource that city administrations must pay to access. This is in spite of the fact that that same data can be and has been generated by applying a set of technologies to a freely available public resource (the street) to predict and accommodate movement. Indeed, the increasing importance of private data sources and operators calls into question the notion of the planned city, as the fulfilment of mobility needs (e.g. the last mile) becomes both liable to failure and subordinate to whether a surplus can be generated from that mobility.

Notes

1 Co-authored with Dr Wen-I Lin at National Taipei University.
2 We make a distinction in this chapter between PBSS1.0 – the older generation of docking and largely municipally governed public bike-sharing – and PBSS2.0, which refers to the newer generation of ICT-based and dockless bike-sharing financed and operated by private firms.

References

Bardhi, F. and Eckhardt, G. (2012) Access-based consumption: The case of car sharing, *Journal of Consumer Research*, 39(4), pp. 881–898.

Behrendt, F. (2016) Why cycling matters for smart cities: Internet of bicycles for intelligent transport, *Journal of Transport Geography*, 56, pp. 157–164.

Bikebiz (2017a) Data mining is why billions are being pumped into dockless bikes. Available at: www.bikebiz.com/data-mining-is-why-billions-are-being-pumped-into-dockless-bikes/ [accessed May 2018].

Bikebiz (2017b) Mobike recognised by UN for environmental impact. Available at: www.bikebiz.com/mobike-recognised-by-un-for-environmental-impact/ [accessed December 2017].

Birchall, C. (2016) Shareveillance: Subjectivity between open and closed data, *Big Data and Society*, pp. 1–12.

Bliss, L. (2018) *Are Dockless Bikes a Cybersecurity Threat?* Available at: www.bloomberg.com/news/articles/2018-02-15/dockless-bikesharing-and-china-s-cybersecurity-laws [accessed May 2019].

Bonham, J. (2006) Transport: Disciplining the body that travels, *Sociological Review*, 54(s1), pp. 57–74.

Borak, M. (2018) *Mobike Co-founder and CEO Davis Wang Resigns, Hu Weiwei Takes the Handlebars*. Available at: https://technode.com/2018/04/28/mobike-davis-wang-resignation/ [accessed May 2019].

Bratton, B.H. (2009) iPhone city, *Architectural Design*, 79(4), pp. 92–97.

Çalışkan, K. and Callon, M. (2009) Economization, part 1: Shifting attention from the economy towards processes of economization, *Economy and Society*, 38(3), pp. 369–398.

China Youth Daily (2017, 6 June) Damaged sharing bikes become new garbage on the streets, the expert argued: irresponsible, *China Youth Daily*. Available at: http://tech.sina.com.cn/i/2017-06-06/doc-ifyfuzny3426519.shtml [accessed March 2019].

Chong, R. (2018, 11 June) Mobike requires no deposit, Ofo declares itself profitable, and Hellobike has raised 2 billion RMB. How can the sharing bike market play in the second wind, *Shanghai Observer*. Available at: www.jfdaily.com/news/detail?id=92689 [accessed March 2019].

Crunchbase (no date) Available at: www.crunchbase.com/organization/mobike [accessed November 2019].

Dafei, Y. (2017) *Big Data Transforms How We Travel*, presentation given at NYU Shanghai Center for Data Science and Analytics, 27 October Available at: https://shanghai.nyu.edu/news/big-data-transforms-how-we-travel [accessed March 2019].

De Souza e Silva, A. (2002) From cyber to hybrid: Mobile technologies as interfaces of hybrid spaces, *Space and Culture*, 3, pp. 261–277.

DeMaio, P. (2004) Will smart bikes succeed as public transportation in the United States?, *Journal of Public Transportation*, 7(2), pp. 1–16.

DeMaio, P. (2009) Bike-sharing: History, impacts, models of provision, and future, *Journal of Public Transportation*, 12(4), pp. 41–56.

DeMaio, P. and Meddin, R. (2017) *Bike Sharing World Map*. Available at: www.google.com/maps/d/viewer?mid=1UxYw9YrwT_R3SGsktJU3D-2GpMU&hl=en&ll=-9.048467847879332%2C138.1453722&z=1 [accessed July 2017].

Deville, J. and van der Velden, L. (2016) 'Seeing the invisible algorithm: The practical politics of tracking the credit trackers', in Amoore, L. and Piotukh, V. (eds) *Algorithmic Life: Calculative devices in the age of big data* (London: Routledge), pp. 87–106.

Dodge, M., Kitchin, R. and Zook, M. (2009) Guest editorial: How does software make space? Exploring some geographical dimensions of pervasive computing and software studies, *Environment and Planning A*, 41, pp. 1283–1293.

Dowling, E. (2019) Valorised but not valued? Affective remuneration, social reproduction and feminist politics beyond the crisis, *British Politics*, 11, pp. 452–468.

Everington, K. (2017) Chinese bike sharing company hits skids after 90% of bicycles lost, *Taiwan News*, 19 June. Available at: www.taiwannews.com.tw/en/news/3191152 [accessed November 2017].

French, M. and Smith, G. (2016) Surveillance and embodiment: dispositifs of capture, *Body and Society*, 22(2), pp. 3–27.

Frith, J. (2013) Turning life into a game: Foursquare, gamification, and personal mobility, *Mobile Media and Communication* 1(2), pp. 248–262.

Gabrys, J. (2014) Programming environments: Environmentality and citizen sensing in the smart city, *Environment and Planning D: Society and Space*, 32(1), pp. 30–48.

Galloway, A.R. (2014) *Laruelle: Against the digital* (Minneapolis, MN: University of Minnesota Press).

Glotz-Richter, M. (2016) Reclaim street space! – exploit the European potential of car sharing, *Transport Research Procedia*, 14, pp. 1296–1304.

Griffin, G.P. and Jiao, J. (2015) Where does bicycling for health happen? Analysing volunteered geographic information through place and plexus, *Journal of Transport and Health*, 2(2), pp. 238–247.

Haixiao, P. (2017) *Air Pollution Crisis Measure and Traffic Management in Shanghai and Beijing* (International Transport Forum). Available at: www.itf-oecd.org/air-pollution-crisis-measures-and-traffic-management-shanghai-and-beijing [accessed February 2018].

Hall, S. and Ince, A. (eds) (2017) *Sharing Economies in Times of Crisis* (Oxford: Routledge).

Harvey, D. (2005) *The New Imperialism* (Oxford: Oxford University Press).

Huang, F. (2018) *The Rise and Fall of China's Cycling Empires*. Available at: https://foreignpolicy.com/2018/12/31/a-billion-bicyclists-can-be-wrong-china-business-bikeshare/ [accessed August 2019].

ITF (2019) *ITF Transport Outlook 2019* (Paris: OECD Publishing). Available at: https://doi.org/10.1787/transp_outlook-en-2019-en.

Jackson, T. (2009) *Prosperity without Growth: The transition to a sustainable economy* (London: Sustainable Development Commission).

Jennings, G. (2015) Finding our balance: Considering the opportunities for public bicycle systems in Cape Town, South Africa, *Research in Transportation Business and Management*, 15, pp. 6–14.

Jessop, B. (2002) *The Future of the Capitalist State* (Oxford: Polity Press / Malden, MA: Blackwell).

Jessop, B. (2006) 'Spatial fixes, temporal fixes and spatio-temporal fixes', in Castree, N. and Gregory, D. (eds) *David Harvey: A critical reader* (Hoboken, NJ: Wiley-Blackwell), pp. 142–166.

Kitchin, R. and Dodge, M. (2011) *Code/Space: Software and everyday life* (Cambridge, MA: Massachussetts Institute of Technology).

Kitchin, R., Lauriault, T. and McArdle, G. (eds) (2018) *Data and the City* (Oxford: Routledge).

Krivy, M. (2018) Towards a critique of cybernetic urbanism: The smart city and the society of control, *Planning Theory*, 17(1), pp. 8–30.

Li, J. (2019) A court has revealed the sorry state of a Chinese bike-sharing startup once worth $2 billion, *Quartz.com*. Available at: https://qz.com/1647637/chinese-bike-sharing-startup-ofo-unable-to-pay-debts/ [accessed August 2019].

Lin, W-I. and Spinney, J. (2020, forthcoming) From public bike to sharing bike: Using mobility as a vehicle to theorise the shift from technology of government to dispositive, *Urban Studies*.

Millington, B. (2014) Smartphone apps and the mobile privatization of health and fitness, *Critical Studies in Media Communication*, 31(5), pp. 479–493.

Mone, G. (2015) The new smart cities, *Communications of the ACM*, 58(7), pp. 20–21.

Ossandon, J. (2013) Low and high finance studies after performativity: A speculative workshop report, *Charisma – Consumer Market Studies*, 8 July. Available at: www.charisma-network.net/markets/low-and-high-finance-studies-after-performativity-a-speculative-workshop-report [accessed 31 October 2014].

Piwek, L., Joinson, A. and Morvan, J. (2015) The use of self-monitoring solutions amongst cyclists: An online survey and empirical study, *Transportation Research Part A: Policy and Practice*, 77, pp. 126–136.

Richardson, L. (2015) Performing the sharing economy, *Geoforum*, 67, pp. 121–129.

Romanillos, G., Zaltz Austwick, M., Ettema, D. and De Kruijf, J. (2015) Big data and cycling, *Transport Reviews*, 1647, pp. 1–20 (January).

Schwanen, T. (2015) Beyond instrument: Smartphone app and sustainable mobility, *European Journal of Transport and Infrastructure Research*, 15(15), pp. 675–690.

Shaheen, S., Guzman, S. and Zhang, H. (2010) *Bikesharing in Europe, the Americas and Asia: Past, Present, and Future* (Davis, CA: UC Davis, Institute of Transportation Studies). Available at: https://escholarship.org/uc/item/79v822k5 [accessed 31 August 2014].

Smith, G. (2016) Surveillance, data and embodiment: On the work of being watched, *Body & Society*, 22(2), pp. 108–139.

Spinney, J. and Lin, W-I. (2018) Are you being shared? Mobility, data and social relations in Shanghai's Public Bike Sharing 2.0 sector, *Applied Mobilities*, 3(1), pp. 66–83.

Star, S.L. (1991) 'The sociology of the invisible: The primacy of work in the writings of Anselm Strauss', in Maines, D.R. (ed.) *Social Organization and Social Process: Essays in honor of Anselm Strauss* (Piscataway, NJ: Transaction Publishers), pp. 265–283.

Stromberg. M. (2017) Ride share, *Planning*, 83(4), pp. 26–27.

Tan, X. and Defei, Y. (2018) *Bike-sharing Data and Cities: Lessons from China's experience*. Available at: www.thegef.org/blog/bike-sharing-data-and-cities-lessons-china%E2%80%99s-experience [accessed May 2019].

Teffer, M. (2018) *Dockless Share Bikes are the Frontline of a Battle between Chinese Tech Giants*. Available at: www.afr.com/lifestyle/cars-bikes-and-boats/cycling/dockless-bikes-are-part-of-a-battle-between-chinese-tech-giants-20180420-h0z1m8 [accessed May 2019].

Tencent News (2017, 6 December) *Mobike awarded the United Nations' highest environmental protection award—Guardian of the Earth Award*. Available at: https://mobike.com/cn/news/460 [accessed March 2019].

The Paper (2018, 1 June) Soliciting opinions of managing sharing bikes in Shanghai: Prohibiting commercial advertisements on bikes, *The Paper*. Available at: http://tech.sina.com.cn/i/2018-06-01/doc-ihcikcew3820748.shtml [accessed March 2019].

UNEP (2017) *UN Environment Honours Seven New Champions of the Earth*. Available at: www.unenvironment.org/news-and-stories/story/un-environment-honours-seven-new-champions-earth [accessed 9 July 2020].

Vallina-Rodriguez, N., Sundaresan, S., Razaghpanah, A., Nithyanand, R., Allman, M., Kreibich, C. and Gill, P. (2016) *Tracking the Trackers: Towards Understanding the Mobile Advertising and Tracking Ecosystem*, Workshop on Data and Algorithmic Transparency (DAT 2016), November, New York.

Vanolo, A. (2014) Smartmentality: The smart city as disciplinary strategy, *Urban Studies*, 51(5), pp. 883–898.

Williamson, B. (2015) Algorithmic skin: Health-tracking technologies, personal analytics and the biopedagogies of digitized health and physical education, *Sport, Education and Society*, 20(1), pp. 133–151.

Xinmin Evening News (2017, 6 March) The Shanghai delegation opened for deliberation yesterday, and Han Zheng answered his views of sharing bikes: 'I highly approve of such an innovative act'. Available at: http://newsxmwb.xinmin.cn/xinminyan/2017/03/07/30884776.html) [accessed March 2019].

Zuboff, S. (2019) *Surveillance Capitalism: The fight for a human future at the new frontier of power* (New York: Hachette Book Group).

8 Conclusions

Where do we go from here?

The four related questions that this book has set out to 'answer' are:

1. What contemporary formations of cycling do we see manifesting in cities?
2. How can we explain the emergence of these formations – why do we see particular formations emerge and not others?
3. How do these formations come into being? Through what processes are social practices of cycling rendered economic?
4. What are the implications of this selective formation for cycling, urbanisation and sustainability?

In these conclusions I provide an overview of the collective chapter insights relating to questions 1–3. I then engage in more detail with the final question to draw out the wider implications in relation to issues around equity, justice and sustainability. I also provide some suggestions for what might be done differently to create a more plural and less economically subordinate vision of cycling.

1 What contemporary formations of cycling do we see manifesting in cities?

The different chapters have highlighted the creation of particular cycling subjects and practices. In line with most cities attempting to recuperate cycling, the most commonly promoted form is commuter cycling, although as the examples of London and Taiwan show, other forms of trick and sporting cycling are also enabled in certain places. However, the examples of London, Shanghai and Taipei show that whether focused on promoting cycling as a form of private or public transport, the focus is largely on facilitating the journey to work. This is echoed in other cities around the world, whether they are cities with high cycle mode share like Amsterdam and Copenhagen or more modest like Seville: success is measured in the number of journeys to work or education, or in the case of PBSS2.0 in Shanghai, by the number of users signed up to use the system.

Even in rare examples that don't involve commuter cycling such as the example of Trials and BMX on London's South Bank, the emphasis is on curating a version of cycling practice that conforms to the wider economic goals of the area as cultural and vibrant. As a result we see different cycling practices managed and confined to specific areas of the city reproducing the modernist tendency to separate street life from the productivity of roads. Ultimately, contemporary formations privilege narrowly conceived productivist (and largely instrumentalised) versions of cycling as appropriate and desirable.

2 How can we explain the emergence of these formations – why do we see particular formations emerge and not others?

Why do we see a narrowly framed version of cycling being promulgated rather than a more plural vision? My main argument across these chapters is that we see these formations because cycling in the contemporary city is seen as crucial to the 'fixing' of a number of related matters of concern that are primariliy economic. The long-term effects of automobilisation on cities are becoming increasingly apparent, with increasing circulation times of goods and labour; the need to invigorate industry and new markets; deteriorating local and global environmental quality; and poorer health outcomes from sedentary lives, particularly automobilised lives. Add to this increasing inter-city competition to be seen as vibrant and liveable on the global stage, and the need for some kind of fix becomes readily apparent if the goal is to sustain economic growth. However, given the political and spatial constraints wrought by the intersections of these different matters of concern, and an overarching neoliberal tendency to shy away from regulation and directly intervene in the economic, the fix in question becomes one that works through the social. In particular, as I have demonstrated here, fixing in the contemporary city increasingly works through the governance of socially reproductive forms of mobility, with cycling – particularly the commute – exemplifying this.

We could argue that the need to address economic matters of concern does not fully explain these economic formations, and indeed they explain it more in a macro political-economic sense. There are, however, key reasons that are closer to the ground, so to speak, with regard to how that politics manifests operationally. The first of these relates to the ethnocentrism evidenced in many professions linked to the promotion and materialisation of cycling. As I have shown, professions such as planning and engineering that are tasked with delivering the cycling agenda are overburdened with middle-aged white men who are much less likely to question a narrowly conceived agenda of commuter cycling because it reflects their 'needs' as male commuters with fewer caring or social reproductive respsonsibilities, social safety concerns, impairments or predilection for less instrumental mobility. That is not to say they cannot question the inclusivity of this agenda, just that given such ethnocentrism, they are much less likely to. In addition to this, the methods of transport appraisal that have evolved reflect economic priorities, and where they don't marginalise cycling completely, they are much better at capturing data on able-bodied and instrumentalised commuter cyclists. It bears repeating that there is a tendency to count what you value and value what you count.

The second related explanation is found in the increasing convergence between physical and virtual mobility due to new technologies. Within urban cycling, innovations in virtual communications have manifested most importantly in the rise of public bike-sharing, both docked but particularly dockless and hybrid systems. Not only have technological developments centred on the smartphone-enabled PBSS2.0 to operate in new, more responsive ways, but the granularity and combination of data they enable is fuelling the creation of new markets that use the behavioural surplus of mobility as the raw material for new products. These hybrid systems of physical and virtual mobility form architectures of extraction that gain the best chance of gathering large datasets on users by facilitating commuter journeys, and thus it is here where their efforts have been most keenly focused.

We should not forget the influence of place on these formations, and in particular logics of inter-city competitiveness. Clearly the examples of Taipei and Taiwan demonstrate that their adoption of cycling has been driven by national economic issues related to their domestic bicycle industry as well as the adoption of a Western framing of cycling as panacea for a variety of urban illnesses. Thus cycling becomes one component in a projected image of Taipei and Taiwan as progressive and liveable in line with other well-known cycling cities and nations. Likewise the accommodation of trick cycling on London's South Bank has been driven by the desire to be seen as liveable in the context of retail-led development and inter-city competition. Place therefore matters, and indeed processes through which places attempt to define and distinguish themselves are bound up with meanings of mobility.

3 How do these formations come into being? Through what processes are social practices of cycling rendered economic?

A key contention across these chapters has been to demonstrate how these formations have been brought into being. Whilst higher-level political-economic concerns alongside technological, geographical and human factors help explain why, they say less about how. The central argument across these chapters is that these formations of cycling emerge due to more or less subtle processes of agencement that seek to bring into being cycling subjects that can – in different ways – generate a behavioural surplus. Here the focus is on the creation of new pro-cycling narratives, material environments, technical objects, ICTs and virtual spaces, policy and legal instruments, corporate–state formations, management strategies, user dispositions, etc. I have shown how through gradual processes of alignment, such elements come together to create users who are knowledgeable, capable and willing to act upon themselves to become cycling subjects. Such an argument takes seriously the fact that user agency is central to the creation of cycling subjects, but it is an agency formatted within loosely strategised arrangements that promote particular qualities of cycling and favour particular outcomes.

As this suggests, contemporary urban cycling agendas reflect the broader neoliberalisation of transport policy where subjects are encouraged to become self-responsible and act upon themselves in a way that harnesses their ability as active agents to assess opportunities, benefits and costs. As a result, the contemporary cycling agenda is one backed by only modest legislative or spatial change, relying instead on a shifting discursive narrative to bring the new cycling subject into being. Such a strategy is, as I have demonstrated, fraught with uncertainty, because in order to become attached to new practices, subjects must become detached from other practices (such as riding the bus or driving). However, with little inducement to detach from these practices, the success of any new attachments is uncertain at best.

I have also adopted and adapted Zuboff's (2019) concept of behavioural surplus alongside ideas of labour power to show how cycling as a social practice is rendered economic. I have argued that by aligning cycling with different matters of concern and technological innovations, the social-reproductive labour of the cyclist can be valorised because labour power exceeds labour time. That is to say, qualities of cycling as a form of mobility that previously held little or negative value are increasingly recognised and harnessed by state and corporate actors to enhance and recuperate processes of capital accumulation.

4 What are the implications of these selective formations for cycling, urbanisation and sustainability?

In this final section I want to broaden out the focus to discuss some of the implications of such a position for mobility, cities and sustainability. As Koglin (2011) has argued, 'much of the sustainability agenda may be questioned and should be seen in the neo-liberal frame, because the term sustainability is so diffuse that everything can be interpreted as sustainable and the economic aspects are often prioritised' (Koglin 2011: 226). Certainly it has been my main contention across these chapters to show that the variants of cycling that are being filtered back into cities are those that facilitate goals of continued economic growth.

The narrow framing of cycling

The first implication is one related to the narrow framing of cycling as mainly focused on commuters that makes it much less likely that certain bodies, ages or types of cycling will occur. The vision of the bike itself (and therefore the rider) projected in urban cycling in countries like the UK is very narrow. Both cycle lanes (in their narrow width and detailed design) as well as the bikes themselves (for example those found in PBSS) privilege the fast, adult, able-bodied commuter due to a near total failure to adequately incorporate non-standard cycles, cargo cycles, those that could accommodate children, and to a lesser extent, those that provide electric assistance. The design of associated infrastructure is an issue in particular, where it allows for fast cycling, but allows little tolerance for less steady and competent cyclists, wider or longer cycles, or more social side-by-side cycling. As a result, a number of users are effectively scripted out of cycling as a practice and cycling becomes inaccessible to many citizens who might otherwise consider it an option.

It is not only the design of infrastructure but the places it connects and disconnects that has an impact on users. As I have shown in the cases of London and Shanghai, cycle infrastructure and PBSS that link residential areas with main employment centres are prioritised, with much less emphasis on everyday mobility and cycling as a form of place-making. The selective placement of cycle infrastructure amounts to a ghettoisation that excludes many groups from cycling because they do not have safe or suitable places to cycle. Braun *et al.* (2019) show just this in a quantitative analysis of bike lanes in 22 cities in the USA. They found that residents on low incomes and from some ethnic groups had significantly lower access to bike lanes than other groups, concluding that bicycle planning and advocacy needed to pay much more attention to issues of social equity.

Similarly, where playful and trick cycling occurs, it is confined and managed into just a few areas of the city that can harness its vitality for economic gain. This suggests a very narrowly framed understanding of liveability, where the unexpected and spontaneous (Southworth 2016) are only allowed in clearly demarcated places and times, rendering them less than spontaneous and effectively another managed attraction in service to retail capital.

We might reasonably question why the priority is given to facilitating commuter journeys given the fact that (in the UK at least) only around one-third of journeys are for this purpose, with the other two-thirds made for leisure, education, and social reproduction. How likely are we to see a meaningful shift to a broader, more inclusive cycling culture if only a minority of journeys for only a minority of citizens are being facilitated? Whilst it may be easier to increase cycling numbers by focusing on citizens who have more straightforward mobility patterns and needs in the context of commuting, I would argue that this agenda may do little in the longer term to replace the car if this remains the most appropriate form of mobility for the other two-thirds of journeys.

The implications for equity and justice are writ large given the focus on commuter cycling, as certain (male, working-age) citizens are given greater mobility choices whilst others (older, younger, female, impaired, ethnic minorities) are not because they are excluded either physically or ontologically. Infrastructure and objects that accommodate commuting qualities of speed, efficiency and directness rather than less instrumental qualities of sociality, playfulness and caring tend to poorly accommodate those with different capabilities or needs who may prioritise these qualities. In particular, narratives, objects and environments that prioritise the lone commuting cyclist in their design and the destinations they serve can exclude those with disabilities, those who do not work, and those with caring responsibilities because cycling remains inaccessible to them as a practice.

Even in a narrowly conceived transportation sense, the impacts in relation to justice and equity are profound, because, as Martens (2017) has argued, a central goal of transportation systems should be to maximise accessibility for as many citizens as possible. With regard to public transportation we have tended to create systems that confer greater levels of accessibility on able-bodied, middle-aged, male commuters because they afford this kind of user through their routing, frequency, physical design and management, with a corollary neglect for the needs of others (Martens 2017; Cahill 2010; Pyer and Tucker 2017; Yavuz and Welch 2009). Rather than learn from these lessons, we seem to be copying the same mistakes in relation to cycling by providing routes that enable fast journeys for the able-bodied and unencumbered to places of work at peak times, but do little to facilitate journeys outside of this that are more likely to be made by women, children, the elderly and impaired. As such, the myopia of vision demonstrated in cycle policy, planning and engineering is an issue of justice and accessibility. In line with Cupples and Ridley, I argue that the overemphasis on journeys to work, and the exclusions it entails, suggests that 'there is an emerging fundamentalism which … might be at odds with promoting social inclusion and might simultaneously fail to apprehend the heterogeneity of environmental responsibility' (Cupples and Ridley 2008: 255).

Transparency and participation

The discussion on PBSS2.0 and user data in particular foregrounds issues of transparency. As the discussion of PBSS2.0 concludes, one of the key issues related to the new wave of bike-sharing is the fact that users have very little idea of what data is being gathered on their behaviour, who is using it, how they are using it or to what ends. Whilst it may be possible to find out given enough work, as Zuboff (2019) has argued, we live in a world where verbose terms and conditions act as an effective deterrent. Moreover, even if a user gains this knowledge and decides they do not want to enter into the exchange, they are excluded from using the service because there is no alternative method of payment except to hand over personal data.

Moreover, whilst PBSS systems undoubtedly can fulfil a transport need, they suffer from an absence of user participation in their planning, design and management, with – in the case of PBSS2.0 particularly – bikes just appearing in streets without local residents being consulted on whether they want such a system, what the system goals should be, what the bikes should look like, where they should be placed in order to maximise participation, or what is being taken as payment. In its present incarnations, PBSS2.0 largely serves commuters with regard to the design of payment systems, bike design and (for hybrid systems) dock locations. If users were meaningfully involved in the physical and virtual design of such systems, and the analysis of data produced, these systems may look very different from those parachuted into cities over the past four years. The importance of such participation and transparency to issues of justice and sustainability should not be underestimated. As Greene and Wegener stated in 1997, any transport system that fails to incorporate meaningful citizen participation cannot be considered sustainable.

As an antidote to the imposition of such systems, Nixon and Schwanen (2019) discuss alternative community-led and sustainable models of bike-sharing that incorporate publics. They cite the example of a community-led bike library in Greater London where the emphasis is on enabling those on low incomes, with little storage and diverse carrying and temporal needs to access bikes. As a result the bikes come in multiple sizes and some have child seats attached. This represents a much more socially embedded (as opposed to economic) relationship with local people and geography, demonstrating, as Nixon and Schwanen point out, 'a deep attunement to the inherent heterogeneity and place specificity of cycling needs and ultimately confronts mobility inequality with respect to income, age, and some degree of body type' (2019: 3).

Cycling: a vehicle for enabling or questioning economic growth?

Building upon this, it is perhaps what current articulations of cycling say about how we conceptualise sustainability that is of most interest. In 1995 Wetzel and Wetzel helped lay the groundwork on the limits to economic growth, citing the need to stay within the carrying capacity of the earth and stating that pro-growth paradigms could not be sustainable: even those with minimal economic growth still required a use of resources and energy that projected into the future would outstrip earth's carrying capacity. More recently, Jackson (2009) has confirmed the folly of believing that anything close to current levels of growth can be maintained given rising levels of affluence, inequalities, and population growth:

> The truth is that there is as yet no credible, socially-just, ecologically-sustainable scenario of continually growing incomes for a world of nine

> billion people. In this context, simplistic assumptions that capitalism's propensity for efficiency will allow us to stabilise the climate or protect against resource scarcity are nothing short of delusional.
>
> (Jackson 2009: 57)

But what does cycling have to do with this? Surely cycling is one of the things that will save us given that it can help us decouple growth from resource use and emissions? A report commissioned by the UCI and ECF in 2015 examined the potential for cycling to reduce emissions and energy use in a high-use scenario, where by 2050 14% of global vehicle kilometres are made by bike (Mason *et al.* 2015). The report suggests that, under such a scenario, an 11% reduction in energy use and emissions is possible by 2050 (2015: 5). Such figures are encouraging and suggest at least a relative decoupling. However, the report also states that even including an aspirational increase in public transport use, emissions from the transport sector would be the same in 2050 as 2015 (5). And though it is hard to verify from the report, it would appear that the calculations do not take into account the fact that there will be 2 billion more people on the earth, all of whom will be wanting a 'developed' standard of living (with a personal carbon budget of around 21 tonnes), a significant increase on the roughly 1.5 tonnes currently emitted each year by the poorest individuals. Even if only the population increase is factored in, overall emissions and energy use would continue to grow despite any decoupling enabled by a switch to cycling.

Again, Jackson's work is insightful on this matter, suggesting that no amount of efficiencies will save us if we remain focused on economic growth as our main goal:

> Those who promote decoupling as an escape route from the dilemma of growth need to take a closer look at the historical evidence – and at the basic arithmetic of growth. Resource efficiency, renewable energy and reductions in material throughput all have a vital role to play in ensuring the sustainability of economic activity. But … it is entirely fanciful to suppose that 'deep' emission and resource cuts can be achieved without confronting the structure of market economies.
>
> (Jackson 2009: 57)

Fletcher and Rammelt (2017) go further, arguing that the emphasis on decoupling represents a delusional neoliberal fantasy that serves to obfuscate the tensions between economic growth, efficiency, population and inequality (450). The danger is that cycling in its current narrow economistic framings has become part of this fantasy narrative and is being used to 'sustain faith in the possibility of attaining sustainable development within the context of a neoliberal capitalist economy that necessitates continual growth to confront inherent contradictions' (2017: 450).

This brings us to the crux of the matter, because current manifestations of urban cycling are not helping us confront the structure of market economies; rather, particular narrowly conceived variants of cycling are being materialised whose success is being measured (and indeed they are only being implemented) because of their ability to enable the continuation of unsustainable growth economies, albeit with (at best) some modest decoupling (Jackson 2009: 53–54). Indeed, as Janicke has stated, patterns of activity that help to sustain high GDP growth 'cannot be "green growth" because it cannot sufficiently be compensated by eco-efficient innovation' (Janicke 2012: 20). As this suggests, a version of cycling that reproduces growth patterns without an underlying transformation cannot be considered sustainable. In helping to reproduce current patterns of growth, current articulations of cycling are, at best, part of a 'weak green economy' (Ferguson 2015: 27) and, at worst, part of a 'green growth' agenda that sees little contradiction between economic growth and sustainability.

From economic growth to happiness and wellbeing

A fundamental issue is that we need to move away from measuring the success of cycling schemes (and in fact all aspects of human existence) in terms of journeys to work (and the economic efficiencies gained), and towards a range of factors focused on wellbeing and happiness. When discussing the effectiveness of different modes of economic organisation in the shift to sustainability, Ferguson notes that our 'conventional pro-growth perspective, assumes that growth is a necessary precondition for societal wellbeing and that a green economy simply involves making existing patterns of growth more resource efficient' (Ferguson 2015: 26). Moreover, Wetzel and Wetzel argue that, 'from the point of view of mainstream economic thought, growth is no longer just an instrument for making people's lives better, but has become an end in itself' (Wetzel and Wetzel 1995: 17). Ferguson argues that it is fundamental to a truly sustainable transformation that we reorient human activity away from measuring economic success and towards measuring other aspects of human flourishing (Cudworth and Hobden 2011, in Ferguson 2015: 26). Whilst many cities around the world have policies that position wellbeing as important, and imply that cycling plays a central role in this, a closer look suggests that it is still a version of cycling whose primary purpose is enhancing economic competitiveness, and indeed any corresponding residual wellbeing that might arise from a shift to cycling is presented as important because it will lower burdens on health services (and therefore make the economy more efficient) rather than as a good in its own right. As Ferguson (2015) has stated, 'the green economy may have the capacity to help animate a transition away from current socially and ecologically unsustainable patterns of economic growth only if ... alternative measures of progress to gross domestic product are widely adopted' (17). As this suggests, if cycling is used as just another way to maximise GDP, it remains part of an unsustainable green growth agenda, rather than helping to transition towards a more sustainable green economy.

Wetzel and Wetzel have argued that 'excessive concern for aggregate growth has migrated from the economics departments of our universities into the culture at large, until it has become the dominant attitude of both private individuals and public officials who shape economic policy' (Wetzel and Wetzel 1995: 178). To this we can include transport policy-makers. Indeed, current articulations of cycling do not suggest any move toward a 'post-growth society' (Ferguson 2015: 18) because they are so focused on facilitating current patterns of growth, rather than letting other intrinsic qualities of cycling be foregrounded that might help produce formations that are more pro-social and less economically oriented.

It is increasingly likely that societal and economic needs for wellbeing and happiness (Layard *et al.* 2008) are not being met through current logics of cycle promotion, which not only seek to sustain capital investment and industrial growth, but distribute what wellbeing gains there are to those of higher socio-economic status. Indeed, cycling in its current incarnation resides at the weaker end of the spectrum of ecological modernisation, characterised as 'economistic, technological, instrumental, narrowly technocratic, corporatist, nationally bounded, and requires no fundamental transformation of the state or economy' (Christoff 1996, in Ferguson 2015: 25). A more strongly ecological variant of cycling would be 'institutionally systemic, communicative, deliberative, democratic, international, and diversifying' (Christoff 1996, in Ferguson 2015: 25).

Does this mean that current manifestations of cycling as part of a green economy serve no purpose? Absolutely not: on the contrary, as Ferguson states they are most likely a necessary precondition on the road to more transformatory modes of economic organisation: 'moving to greener forms of growth, even if they merely constitute relative decoupling, appears to be the only feasible first step towards a post-growth economy' (Ferguson 2015: 22). Perhaps the best thing we can say about current manifestations of urban cycling is that, positioned within discourses of green economy, they avoid expressing an explicit 'anti-growth' position that, as Barry (2012) has argued, often hold environmental politics back because of a direct challenge to dominant pro-growth discourses that are seen as too radical (cited in Ferguson 2015: 29). However, we cannot hope that cycling policy will somehow inevitably move toward a stronger articulation of green economy as part of a fait accompli; we must actively move it in that direction by highlighting the shortcomings of current incarnations.

Cycling is dead, long live cycling: toward place-making, participation and inclusivity

Place-making

If we imagine cycling as only a practice that can enhance economic competitiveness through the creation and valorisation of human capital, etc., we reproduce the city as a primarily economic entity. Some might argue this is a fait accompli; after all, the expansion of the modern city can be traced back to the needs of industrialists and therefore its spatial arrangements and the subjects enabled within tend to privilege those that best serve the economy. However, whilst human flourishing requires some form of labour to be undertaken, such a fatalistic reading forgets that human flourishing is about so much more than just labour. Indeed, such narrowly defined ideas of human flourishing tend to marginalise the importance of social interactivity, identity creation, community formation, emotional engagement and playfulness to wellbeing.

The late Doreen Massey (2004) argued that because we all have a say in how to produce and consume (and move around), an alternative city is never far away from being achieved through our roles as individual and collective decision-makers. She goes on to call for citizens to be active in liberating the 'urban imagination from the false demands and constrictions of neo-liberal globalisation and from the city dominated by pervasive capitalism'. As Harvey has stated:

> We must imagine a more inclusive, even if continuously fractious, city based not only upon a different ordering of rights but upon different political-economic practices. If our urban world has been imagined and made, then it can be re-imagined and re-made.
>
> (Harvey 2003: 941)

We must include our mobility systems in this, which cannot only be shaped to serve the economy and alleviate matters of concern that are ostensibly economic: these systems should reflect a broader set of needs and desires and a broader range of social and cultural identities that are 'unproductive' in their intent but central to wellbeing.

Participation

It is clear from the chapters in this book that cycling policy and practice, from planning and engineering through to the corporate architects of PBSS2.0, has a problem with citizen participation. Fotel suggests that elites still control the idea of what mobility should be, with the result that many people are pushed around and have their lives defined by others' mobility, particularly through the unjust production of spaces through cynical urban planning (Fotel 2006: 733, 743–744). The results are bicycle systems that construct and reflect the needs of a few, not the many; and which fail to engage citizens in the design of systems or collection and analysis of data. Designed with narrowly framed policy goals in mind, contemporary cycling mobilities are highly passive and fit Barr's description of smart mobilities as 'disengaged forms of practice, which do not offer the political possibility of considering more radical alternatives to the mantra of atomistic and incremental behavioral change' (Barr 2018: 7).

A key issue here is the need to recognise more qualities of cycling as important (rather than just those that address top–down identified matters of concern) in order to bring into being cities and society that prioritise humanity rather than economy. A good start here would be to recognise which qualities of mobility are valued by different ages, genders and ethnicities, and seek to accommodate these (Handy and Lee 2020). Whilst some currently promoted qualities will no doubt remain important, it is likely that others will also rise to prominence, which would suggest different legislative frameworks, spatial arrangements, materialities and agencies. It may effectively produce new ontologies of cycling that are less subservient to prevailing economic agendas and better reflect multiple readings of cycling as a social practice.

In order to achieve this we need to change the voices that we are hearing in cycling debates. Amsterdam's recent appointment of a junior cycling mayor to represent children's interests in the city is an example of a step in the right direction. However, rather than rely on representation, we need to incorporate the voices of children (and other potential cyclists) into decision-making in more fundamental ways. Methodologies and engagement strategies must be developed that allow othered elements of cycling mobilities to come to the fore, and that create participatory spaces where local power dynamics can be addressed and othered voices can be heard and acted upon (Johnson 2017: 105).

Reflexivity and place-making

The mobilities turn has amply demonstrated that the meanings of mobility cannot be reduced to getting from A to B (Cresswell 2006; Sheller 2004; Ferguson 2009) and that wellbeing is central to journey experience (Nordbakke and Schwanen 2014). However, a mobility system that enables only certain journey qualities (e.g. efficiency) runs the risk of marginalising other important qualities that contribute to a more holistic and rounded notion of wellbeing. Accordingly, I argue here that if different qualities of cycling were foregrounded, it could offer so much more as a vehicle to transform the norms of industrial society, allowing us, as Beck (1992) has argued, to renegotiate our 'relationship with nature radically, as the social practices that have traditionally shaped industrialisation' become more fluid and amenable to change (cited in Ferguson 2015: 24). As Ferguson goes on, 'the transformative potential of strong ecological modernisation stems from its "reflexivity" (Paterson *et al.* 2006): it encourages continual reflection upon the consequences and purposes of the modernisation process and the means and ends of social, political, and economic action' (Ferguson 2015: 25).

Barr has argued that roads 'should be repurposed as ones for socializing and dwelling, rather than merely places through which we have to move to serve an economic purpose' (Barr 2018: 8) As Barr goes on to note, we need to recuperate the notion that 'dwelling, rather than constantly moving, might be an aspiration, and one which lays at the heart of efforts to sustain the vibrancy of urban areas' (Barr 2018: 8). I would suggest that cycling – if more variants of it and more practitioners of it are accommodated – has the potential to bring about such a shift. Alternative qualities such as playfulness, sociality, and slowness offer an implicit critique of the instrumentalism embedded in current city form and mobility. Rather than exclude these, we need to welcome them in to enable a reflexiveness to emerge that changes what we value. Such a shift will not directly enhance economic productivity, but it will likely enhance wellbeing, and for a greater number of inhabitants. Doing so gives us the chance to move cities away from being engines of growth to being engines of wellbeing. However, if we continue to foreground qualities such as speed and instrumentalism that champion the existing norms of industrial society, this reflexive potential is unlikely to be realised.

Accordingly, I would argue that cycling policy needs to forget its fixation on commuting and efficiency, and focus instead on planning cities around a more multiple reading of cycling and a broader cast of users. This requires looking in a more ethnographic way at what cyclists actually do (rather than a selective reading of the bits that are 'useful') and accommodating these qualities and practices at the neighbourhood level rather than trying to fit them in at the margins. To make this a reality will entail bold and unpopular decisions relating to car use; require wholesale mixing of land uses; require bringing agriculture, nature and industry into cities. In short, it is a more radical vision of Kenworthy's (2006) eco-city that seeks to disable economic growth in favour of social flourishing.

Escaping the silo

Finally, I would argue that we need to get out of current silos in mobility thinking and go beyond ideas of discreet modes. Can cycling – in its currently individualistically oriented variants – possibly accommodate all the users it needs to, or do we need a new and less clearly defined form of active travel that includes everything from the roller skate to the quadricycle, and a whole host of things we have yet to even imagine? Such a move starts to include many different user groups because it enables multiple performances: children to move in the (playful) ways they want; those with caring responsibilities to be able to fulfil these roles; those with impairments to transcend disabling environments and objects, etc.

In the same way that Lorimer and Marshall (2015) have suggested we need to go beyond cycling and walking as the only forms of active travel in town, we also need to cut across the car–bike divide if we are to open active mobility up to a greater range of citizens. Current binary logic pits the bike against the car, when perhaps what is needed is a solution that hybridises the two. Electric cars and electric bicycles only go so far in creating more sustainable modes of transport and evidently can't accommodate the needs of all users or address all matters of concern. However, a more hybrid solution, such as the ped-elec quadricycle (cf. Vital Garcia-Verdugo 2012) could just offer enough of the comfort, care, accessibility and sociality articulated through the car with the health, environmental, place-making and fun benefits of the bike to be a greater step towards sustainability. Whilst this book has not even engaged with freight transport, clearly an understanding of the bicycle as something more than a two-wheeled upright machine would also facilitate towards active and more sustainable freight modes. It is time that we stopped putting cars and bikes on different ends of the spectrum and thought more open-mindedly about what else we can put on our roads.

By way of final conclusion, I first want to reiterate a key message of the book: I am not arguing that the popularisation of cycling is inherently bad; rather, it is the manner in which cycling is being popularised – bringing a narrowly defined version of cycling back into cities – that should give us cause for concern. Ultimately this book argues that cycling has now become a vehicle for sustaining pro-growth agendas rather than subverting them or shifting them in favour of more sustainable no-growth, de-growth, wellbeing-focused and less technologically driven visions of late modernity. Indeed, there are many small-scale community bike initiatives that may deliver better 'value' in the form of community empowerment, playful and 'slow' mobility, and greater accessibility, yet these are not the ones being fostered (Nixon and Schwanen 2019).

Whilst the focus remains on instrumental movement whose success is measured with respect to its contribution to economic growth (whether that be faster circulation times, more desirable retail destinations, more productive human capital, the creation of predictive products, or a more globally competitive image), cycling will remain central to the game of continued unsustainable economic growth. Only when other qualities of it are foregrounded – its slowness, its sociality, its openness, its playfulness – and its success is measured in relation to its role in place-making rather than instrumental mobility, might urban cycling enable us to move beyond our current unsustainable growth paradigm.

Thus it is not just how we transport ourselves, but the goals to which that transport is directed that dictate its sustainability. In the same way that increasing access times and greening journeys to airports through the provision of public transport is fundamentally unsustainable, so too is using cycling to speed up circulation times to maintain current patterns of production and consumption: whilst cycling may help to decouple transport from emissions, it continues to feed growth rather than question it. Even if the switch to a cycle journey does improve individual wellbeing, if that cycle journey helps to fuel continued and unsustainable economic growth, in the long run it will contribute to a reduction in collective wellbeing. So it is not only how we travel that matters, but the goals to which that travel is oriented that matter. Contemporary cycling policy urgently needs to recognise this.

References

Barr, S. (2018) Personal mobility and climate change, *Advanced Review*, 9(5), pp. 1–19.

Barry, J. (2012) *The Politics of Actually Existing Unsustainability: Human flourishing in a climate-changed, carbon-constrained world* (Oxford: Oxford University Press).

Beck, U. (1992) *Risk Society: Towards a new modernity* (London: Sage).

Braun, L., Rodriguez, D. and Gordon-Larsen, P. (2019) Social (in)equity in access to cycling infrastructure: Cross-sectional associations between bike lanes and area-level sociodemographic characteristics in 22 large U.S. cities, *Journal of Transport Geography*, 80, pp. 1–10.

Cahill, M. (2010) *Transport, Environment and Society* (Maidenhead: Open University Press).

Christoff, P. (1996) Ecological modernisation, ecological modernities, *Environmental Politics*, 5(3), pp. 476–500. doi:10.1080/09644019608414283.

Cresswell, T. (2006) *On the Move* (Oxford: Routledge).

Cudworth, E. and Hobden, S. (2011) Beyond environmental security: Complex systems, multiple inequalities and environmental risks, *Environmental Politics*, 20(1), pp. 42–59. doi:10.1080/09644016.2011.538165.

Cupples, J. and Ridley, E. (2008) Towards a heterogeneous environmental responsibility: Sustainability and cycling fundamentalism, *AREA*, 40(2), pp. 254–264.

Ferguson, H. (2009) Driven to care: The car, automobility and social work, *Mobilities*, 4(2), pp. 275–293.

Ferguson, P. (2015) The green economy agenda: Business as usual or transformational discourse?, *Environmental Politics*, 24(1), pp. 17–37.

Fletcher, R. and Rammelt, C. (2017) Decoupling: a key fantasy of the post-2015 Sustainable Development Agenda, *Globalizations*, 14(3), pp. 450-467, doi: 10.1080/14747731.2016.1263077.

Fotel, T. (2006) Space, power and mobility: Car traffic as a controversial issue in neighbourhood regeneration, *Environment and Planning A*, 38, pp. 733–748.

Greene, D. and Wegener, M. (1997) Sustainable transport, *Journal of Transport Geography*, 5(3), pp. 177–190.

Handy, S. and Lee, A. (2020) What is it about bicycling? Evidence from Davis, California, *Travel Behaviour and Society*, 20, pp. 348–357.

Harvey, D. (2003) The right to the city, *International Journal for Urban and Regional Research*, 27, pp. 939–941.

Jackson, T. (2009) *Prosperity without Growth: The transition to a sustainable economy* (London: Sustainable Development Commission).

Janicke, M. (2012) Green growth: From a growing eco-industry to economic sustainability, *Energy Policy*, 48, pp. 13–21, doi:10.1016/j.enpol.2012.04.045.

Johnson, V. (2017) Moving beyond voice in children and young people's participation, *Action Research*, 15(1), pp. 104–124.

Kenworthy, J. (2006) The eco-city: Ten key transport and planning dimensions for sustainable city development, *Environment and Urbanisation*, 18(1), pp. 67–85.

Koglin, T. (2011) Planning for cycling = planning for equity: A response to Cupples and Ridley 'Towards a heterogeneous environmental responsibility: sustainability and cycling fundamentalism' (2008), *Area*, 43(2), pp. 225–227.

Layard, R., Mayraz, G. and Nickell, S. (2008) The marginal utility of income, *Journal of Public Economics*, 92(8–9), pp. 1846–1857. doi:10.1016/j.jpubeco.2008.01.007.

Lorimer, S. and Marshall, S. (2015) Beyond walking and cycling: Scoping small-wheel modes, *Proceedings of the Institution of Civil Engineers – Engineering Sustainability*, 169(2), pp. 58–66.

Martens, K. (2017) *Transport Justice* (New York: Routledge).

Mason, J., Fulton, L. and McDonald, Z. (2015) *A Global High Shift Cycling Scenario* (Davis, CA: Institute for Transportation and Development Policy, University of California, Davis).

Massey, D. (2004) Geographies of responsibility, *Geografiska Annaler B: Human Geography*, 86, pp. 5–18.

Nixon, D. and Schwanen, T. (2019) Bike sharing beyond the norm, *Journal of Transport Geography*, 80, pp. 1–4.

Nordbakke, S. and Schwanen, T. (2014) Well-being and mobility: A theoretical framework and literature review focusing on older people, *Mobilities*, 9(1), pp. 104–129.

Pyer, M. and Tucker, F. (2017) 'With us, we, like, physically can't': Transport, mobility and the leisure experiences of teenage wheelchair users, *Mobilities*, 12(1), pp. 36–52, doi:10.1080/17450101.2014.970390.

Sheller, M. (2004) Automotive emotions: Feeling the car, *Theory, Culture and Society*, 21,(4–5), pp. 221–242.

Southworth, M. (2016) Learning to make liveable cities, *Journal of Urban Design*, 21(5), pp. 570–573, doi:10.1080/13574809.2016.1220152.

Vital Garcia-Verdugo, L. (2012) *uCar: Multi-disciplinary development of an electric vehicle typology for the city*, unpublished PhD thesis (Royal College of Art, London). Available at: http://researchonline.rca.ac.uk/1354/1/VITAL%20Lino%20Thesis.pdf.

Wetzel, K. and Wetzel, J. (1995) Sizing the earth: Recognition of economic carrying capacity, *Ecological Economics*, 12, pp. 13–21.

Yavuz, N. and Welch E. (2009) Addressing fear of crime in public space: Gender differences in reaction to safety measures in train transit, *Urban Studies*, 47(12), pp. 2491–2515.

Zuboff, S. (2019) *Surveillance Capitalism: The fight for a human future at the new frontier of power* (New York: Hachette Book Group).

Index

Page numbers in *italics* denote figures.

For Product Safety Concerns and Information please contact our EU representative GPSR@taylorandfrancis.com
Taylor & Francis Verlag GmbH, Kaufingerstraße 24, 80331 München, Germany

www.ingramcontent.com/pod-product-compliance
Ingram Content Group UK Ltd.
Pitfield, Milton Keynes, MK11 3LW, UK
UKHW021721240926
475702UK00029BA/1746
9780367567736